Power Systems

Electrical power has been the technological foundation of industrial societies for many years. Although the systems designed to provide and apply electrical energy have reached a high degree of maturity, unforeseen problems are constantly encountered, necessitating the design of more efficient and reliable systems based on novel technologies. The book series Power Systems is aimed at providing detailed, accurate and sound technical information about these new developments in electrical power engineering. It includes topics on power generation, storage and transmission as well as electrical machines. The monographs and advanced textbooks in this series address researchers, lecturers, industrial engineers and senior students in electrical engineering.

Power Systems is indexed in Scopus

More information about this series at http://www.springer.com/series/4622

Ryszard Bartnik · Tomasz Wojciech Kowalczyk

Hierarchical Gas-Gas Systems

Thermal and Economic Effectiveness

Springer

Ryszard Bartnik
Department of Power Engineering
and Management
Opole University of Technology
Opole, Poland

Tomasz Wojciech Kowalczyk
Institute of Fluid Flow Machinery
Polish Academy of Sciences
Gdansk, Poland

ISSN 1612-1287 ISSN 1860-4676 (electronic)
Power Systems
ISBN 978-3-030-69204-9 ISBN 978-3-030-69205-6 (eBook)
https://doi.org/10.1007/978-3-030-69205-6

This Springer imprint is published by the registered company Springer Nature Switzerland AG
The registered company address is: Gewerbestrasse 11, 6330 Cham, Switzerland

Contents

Notations

a, b, c, d, e, x	Auxiliary quantities
A, B	Auxiliary quantities
$\dot{B}$	Exergy stream
c_p	Specific heat capacity at constant pressure
$\dot{C}$	Heat capacity stream
e	Unit prices
$\dot{E}_{ch}$	Fuel chemical energy stream
i	Specific enthalpy
i	Unit investment outlays
$\dot{I}$	Enthalpy stream
J	Investment outlays
k_c	Unit heat production cost
k_{el}	Unit electricity production cost
k_z	Unit cooling production cost
L	Mechanical work
$\dot{m}$	Mass stream
N	Power
NPV	Net updated value
p	Pressure
p	Unit rates for environmental emission
p	Income tax rate for gross profit
$\dot{P}_{\text{sup}}$	Afterburn fuel stream (supplementary firing)
$\dot{Q}$	Heat stream
Q_R	Heat and cooling production per annum
q	Fuel afterburn rate
r	Capital interest rate
s	Entropy
$\dot{S}$	Entropy stream
T	Power plant, combined heat and power plantor chiller operating period expressed in years
T	Absolute temperature
$\bar{T}$	Entropic average temperature

W_d	Heating value
z	Compression ratio
z	Investment capital freeze rate
β, ξ	Auxiliary quantities
δ, Δ	Increment symbol
δ_{serv}	Annual fixed cost rate depending on investment outlays (costs of maintenance and repairs of devices)
ε_{el}	Relative rate for the own needs of electricity
ε	Chiller efficiency
η	Efficiency
k	Isentropic and circulation medium exponent
ρ	Environmental emission
σ	Cogeneration rate
τ	Time

Chapter 1
Introduction

For the things of this world cannot be made known without a knowledge of mathematics
(Roger Bacon, 1214–1294)

Hierarchical systems (Figs. 2.3, 2.4, 3.1, 3.2 and 3.3) are multi-level, clockwise or anticlockwise systems. The clockwise systems include heat engines, while the anticlockwise—heat and power machines. The basic property of hierarchical systems is the fact that heat from an external heat source is input only for one cycle. For an engine it is the cycle superior in hierarchy, i.e. the cycle operating at the highest temperature range, while for a heat and power machine, chiller or heat exchanger it is the lowest cycle in the hierarchy, i.e. the cycle operating at the lowest temperature range. For each remaining cycles the input heat is the heat exported from cycles, in the case of engine, in the hierarchy directly above them, while for a heat and power machine, from cycles located directly below them (Figs. 2.3 and 2.4).

Most importantly, in hierarchical engines the efficiency of fuel chemical energy conversion into mechanical work is significantly higher than obtained for single-cycle engines. This is because a significantly higher temperature range is used in these systems from the range $\langle T_g; T_{amb}\rangle$, i.e. from the range between higher heat source temperature T_g and ambient temperature T_{amb} that makes the lower heat source. In the most thermodynamically ideal, theoretical *Carnot* engine with a theoretically maximum possible power the whole $\langle T_g; T_{amb}\rangle$ range is utilised. The power of *Carnot* engine is illustrated by a rectangle area drawn with a broken line in Fig. 1.1. Therefore, the higher the number of cycles j in the hierarchical system—Fig. 2.3—with different temperature operating ranges, the more the system will use the $\langle T_g; T_{amb}\rangle$ range. Thus, the lower will be the loss in exergy stream and the higher will be its power (Sect. 2.2). Within the limits, when the difference $j \to \infty$ between the efficiency of the theoretical *Carnot* engine (2.5) and the efficiency of a hierarchical engine η_{1-j} (2.22) disappears, $\eta_C - \eta_{1-j} \to 0$, and the mechanical power of the hierarchical engine equals the power of the theoretical *Carnot* engine (2.4), i.e. the maximum power available thanks to the use, as for *Carnot* engine, the whole temperature range $\langle T_g; T_{amb}\rangle$.

R. Bartnik and T. W. Kowalczyk, *Hierarchical Gas-Gas Systems*,
Power Systems, https://doi.org/10.1007/978-3-030-69205-6_1

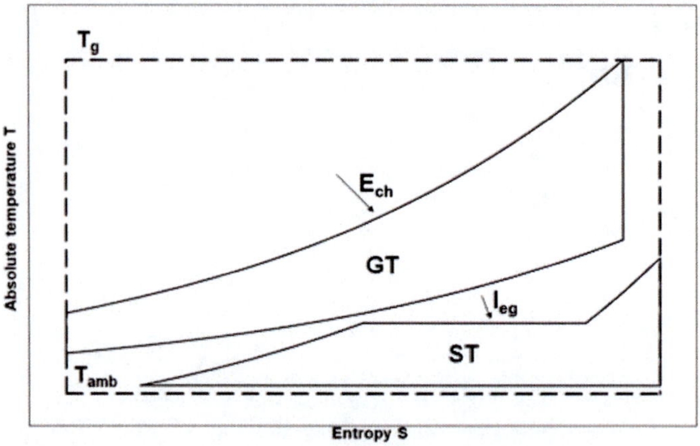

Fig. 1.1 Comparative (theoretical) cycle of a hierarchical gas-steam system (GT—gas turbine Joule-Brayton cycle, ST—steam turbine Clausius-Rankine cycle, E_{ch}—fuel chemical energy supplied to GT, I_{eg}—enthalpy of exhaust gases from the gas turbine supplied to ST by using a heat recovery steam generator; dotted lines show the Carnot cycle for extreme temperatures T_g, T_{amb})

Completely opposite situation can be encountered for the power and heat machine, chiller or heat pump (Sect. 2.3 and Fig. 2.4). Elevating the number of cycles used over one increases the exergy loss, thus boosting the machine drive power. The theoretically lowest drive power features the machine with the anticlockwise *Carnot* cycle.

Currently, in real-world applications hierarchical dual-cycle *gas-steam* engines are used [1]—Fig. 1.1.

They are engines that currently feature the highest efficiency of converting fuel chemical energy into mechanical work and, consequently, as per *Faraday*'s law in an electricity generator, the most noble and precious energy form in all meaning of the word. That is because the widest temperature range of $\langle T_g;\ T_{amb}\rangle$ has so far been used for them. At the high temperature range of the gas turbine, the *Joule-Brayton* cycle is used, while at the low temperature range of the steam turbine, the *Clausius-Rankine* cycle is used. Combining the *Joule-Brayton* cycle and the *Clausius-Rankine* cycle into a hierarchical system is carried out via the exhaust *gas-steam-water* system in the heat recovery steam generator (HRSG), where the low-temperature enthalpy of exhaust gases from the gas turbine is used to generate steam supplying the steam turbine [1]. The power of the *gas-steam* engine is expressed by the sum of TG and TP surface areas presented in Fig. 1.1. The sum of those surface areas is significantly lower than the surface area for the *Carnot* cycle, which, as already stated above, is the surface area of the rectangle drawn with a dotted line in the figure. The net efficiency of hierarchical *gas-steam* engines already exceeds even the value of 60% (efficiency for the theoretical Carnot engine (2.5)), for the same T_g temperature, as the temperature of exhaust gases supplied to the gas turbine—Fig. 1.1—is approx. 85%. Here it is worth noting that the pressure of steam in *gas-steam* hierarchical engines is only approx. 5.5–10 MPa and not like e.g. 28 MPa in units for supercritical fresh steam where only the *Clausius-Rankine* cycle is used. The gross efficiency for electricity generation in these supercritical *Clausius-Rankine* cycles is up to 50%, approx. 46% net. However, the units have already achieved their top development. Only the supercritical parameters (fresh steam temperature and pressure would have to be up to 720 °C and 35 MPa) could increase the efficiency. The increase would be about by a few percentage points. So far there has been no steel available that could "bear" such high parameters. However, exhausting all the opportunities for developing supercritical technology is not an allegation. Just the opposite, it indicates that the technology has reached its maximum performance.

The basic limitation for electricity generation in mechanical engines is that, according to the 2nd Law of Thermodynamics (one of its formulations is formula (2.5)), heat cannot be converted into work in 100%. When from the chain of thermodynamic conversions that take place in the process of electricity generation, the conversions of changing heat into mechanical work, so when the chemical energy of fuel is converted into electric energy, i.e. by omitting "the *Carnot* corset", then the theoretical efficiency for electricity generation in this case is 100%. This applies to fuel cells, whose current efficiency can at present exceed 70%. However, in the case of using hydrogen as fuel in fuel cells the efficiency should include the power efficiency of its generation, which in life cycle cost analysis will significantly reduce the efficiency of cells. Therefore, this is very important, which is the case, to further develop their technology to obtain even higher efficiency, because the range from 70 to 100% is still quite wide. Another significant advantage of fuel cells is their small size as compared to mechanical engines featuring the same electric power. Considering the above-mentioned, chapter 5 presents the thermodynamic and economic analysis of hydrogen production in the process of water electrolysis in the system with a hierarchical *gas–gas* engine—see Fig. 5.1—and for comparison with all remaining currently available power generation technologies.

In addition, it is also possible to use mechanical *gas-steam* quasi-hierarchical engines, i.e. dual-fuel engines, where as opposed to hierarchical engines, the fuel is supplied to both cycles. The *Joule-Brayton* cycle is supplied with natural gas, while the *Clausius-Rankinecycle* is supplied with coal [2–4]. The gross efficiency of dual-fuel engines exceeds even 50% while a steam part operates, which is significant, with subcritical fresh steam parameters, i.e. this is comparable to the efficiencies of the above-mentioned steam units using only the *Clausius-Rankine* cycle for supercritical conditions. The level of efficiency for quasi-hierarchical engines grows to 50% along with the increase in power used in the gas turbine system [2–4]. Significantly, investment outlays per unit (per unit of electric power installed) for those engines are significantly lower than the outlays on units using supercritical parameters. However, the efficiency of quasi-hierarchical *gas-steam* engines is still obviously lower than the efficiency of hierarchical *gas-steam* engines that, additionally, require even lower unit outlays. However, it should be noted that dual-fuel *gas-steam* units can be created by adding the gas turbine to already existing coal-fired power units using *Clausius-Rankine* cycle [2–4]. This solution provides a huge upgrade potential for domestic power industry using mainly, which is justified, coal. Because coal guarantees Poland's power independence and safety. At the same time, it is necessary to build nuclear power units in Poland, which is reasonable for technological, environmental and economic reasons [5]. The unit cost of electricity generation for its operating life ranging from 40 to 60 years (currently 60 years is normal life cycle period for nuclear units) and for low-interest rate investment outlay is small—Fig. 5.2. At the same time, we should develop the fuel element manufacturing industry. The resources of uranium in Poland are high.

This monograph presents the thermodynamic and economic analysis of hierarchical *gas–gas* systems. The *gas–gas* engine uses mutually combined two *clockwise Joule-Brayton* cycles. Fuel is supplied only to the *Joule-Brayton* gas turbine operating at the high temperature range. To the second of the *Joule-Brayton* cycle, i.e. to the cycle used in the turboexpander operating at the low temperature range, the input drive heat is the heat output from the gas turbine *Joule-Brayton* cycle so from the cycle being immediately higher in the hierarchy than it is. Combining both cycles to create a hierarchical system takes place in the air heater, where exhaust gases discharged from the gas turbine heat the compressed air that supplies the turboexpander (Figs. 3.1, 3.2 and 3.3). However, the efficiency η^{G-G} ((3.13), Figs. 3.10 and 4.23) of electricity generation in the *gas–gas* system is lower than in the *gas-steam* system. The superior efficiency of the *gas-steam* engine depends on the steam condensation isotherm when heat from the *Clausius-Rankine* cycle is discharged to the atmosphere. This isotherm almost matches the ambience isotherm T_{ot}—Fig. 1.1—of the *Carnotengine*, the engine, as already mentioned, being the most ideal in terms of thermodynamics. By contrast, the heat transfer medium in the isobaric process of "lower" *Joule-Brayton* cycle in the hierarchical *gas–gas* engine when heat is discharged from the engine to the atmosphere has a significantly higher T_{ot} entropic average temperature (Chap. 2). However, it is important that financial investment outlays on the *gas–gas* engine make only 45% of the outlays on the *gas-steam* engine. After all, this is the economic viability of the engine operation that affects taking the investment decision to build it (Sect. 3.3).

Another, exceptionally important matter in this case. The efficiency of *gas–gas* η^{G-G} engines (3.13), for contemporary gas turbines, i.e. for high temperatures T_{ot} (Figs. 3.10 and 4.23), is comparable to the efficiency of steam engines using *Clausius-Rankine* cycle for fresh steam supercritical parameters. Unit financial outlays on *gas–gas* engines are, at the same time, which is extremely important, a few times lower as compared to the outlays on *Clausius-Rankine* cycle "supercritical" engines. Therefore, we can expect that *gas–gas* engines to be more economically viable despite the fact that natural gas used to fire them is over twice more expensive per chemical fuel energy than coal.

The *gas–gas* engine can also drive, e.g. a heat and power machine—Figs. 2.5 and 4.1a, b. A heat and power machine, compressor chiller or a compressor heat pump use, obviously, the anticlockwise *Joule-Brayton* cycle. Combining subsequent cycles (stages) of a chiller or heat pump, lower in the hierarchy just above them, takes place in heat exchangers operating at the same time as condensers for lower cycles and evaporators for higher cycles—Fig. 2.4. For comparative purposes, the monograph also includes the analysis of the heat chiller, i.e. the chiller driven by mechanical work, not powered by heat, the low-temperature enthalpy of exhaust gases from the *gas–gas* engine—Fig. 4.1c.

It should be noted that the chiller and heat pump use different operation temperature ranges. For the chiller the input heat is the heat discharged from the chilling chamber with a temperature lower than ambient temperature. Then, the heat is discharged to the atmosphere. By contrast, the heating heat input to the heat pump is the heat taken from the atmosphere and supplied into the heated space with temperature higher than ambient temperature (therefore, the cycles of the chiller and heat pump can be combined as the heat flow directions inside them between heat transfer media and the atmosphere are opposite). However, it should be noted that the compressor heat pump under Polish conditions, but not only, is not viable in terms of thermodynamics. The heating heat taken by the pump from the atmosphere is more than a dozen percent lower from the heat input to it from power plants generating electricity used to drive it—Fig. 2.7. This is due to low power efficiency of domestic power plants. The compressor heat pump is especially economically unviable [6]. The heating heat acquired from this pump would have to be subsidised, because it is very expensive. It results from a high annual cost of electricity used to power it and high investment outlays on it, especially for installation-construction work, mainly for installing the evaporator in the soil at a depth of approx 1.5 m on the surface of at least 500 m^2 for a single-family house. Furthermore, the use of heat pumps, especially hierarchical ones, does not make any technological sense, as the heating heat can be acquired directly from working engines in combination with electric power generation [6]. The heating heat acquired in this way is the cheapest. Comparing to the heat from the heat pump it is many times cheaper. Because of the above-mentioned this monograph does not practically describe the heat pump, apart from the short Sect. 2.5 (the thermodynamic and economic analysis of the heat pump is presented in [6]). However, obviously, the methodology of exergy and energy balance for *j*-cycle compressor heat pump is the same as for the *j*-cycle compressor chiller. Only their different temperature operating ranges, i.e. contrary purposes they are used to (chilling and heating), result in changing some characters from + to − in the balance of entropy, exergy and energy, because of different sources of inputting and outputting heat from both machines (formulas (2.4), (2.5), (2.6); see also Sects. 2.4 and 2.5).

Currently in engineering and sometimes scientific practice (however it should not take place), ready-made, numerous commercial applications for performing simulation calculations, e.g. Gate Cycle™ are used (it should be emphasised that the applications are "black boxes" for the user). The results of power analyses obtained this way are just numerical results acquired for specific numerical input data. Extensive experience and knowledge of thermodynamics of problems being solved is needed to use the applications. Entering calculation input data requires knowledge of optimum solutions. Otherwise, the results obtained, proper in terms of balances, are far from being optimum, if not just incorrect. Furthermore, too much simultaneous input data needed for calculation and related interactions do not allow us to generalise the considerations by using detailed numerical results, the only correct way is from a general view to a detail and such an approach makes it possible. By contrast, going from a detail to a general view –not to say normally—is not true (however, it should be made clear that to conceive the truth and formulate it we need to perform a lot of experiments, calculations and detailed analyses and in the first place, we need a smart mind; otherwise, it would be hard to notice the truth about the whole physical processes, phenomena and not only). It should be also emphasised that using the applications does not make it possible to find the answer to the almost fundamental questions: which of the hierarchical systems *gas–gas* or *gas-steam* is more economically and energetically more effective? Therefore, only an analysis using own mathematical models, own calculation codes as presented in the monograph, allows us to answer the question. Analytical own models are also precious, because they allow us to obtain a lot of additional, important information on properties and the properties of *gas–gas* systems under consideration. They allow us *explicite* to assess the impact of individual input values on final results and in the first place easily and quickly find not only an optimum solution, but also a range of solutions close to optimum; in addition, they make it possible to show the nature of changes. They allow us to discuss and analyse the results. For technological purposes it shows a big, relevant value. What is more, the analytical models will allow us to draw conclusions, which is extremely important, and as described above, general in its nature. Solutions obtained by using the economic criterion in the analysis enable us additionally to draw conclusions on economic requirements for implementing a new power generation technology allowing us to specify economically justified relations and the range of energy media prices. The methodology of technical–economic analyses of *gas–gas* systems presented in the monograph has both cognitive values and extends the knowledge about these systems as well as provides a wide range of application operations. However, in order to make it possible to develop these models, it was necessary to make simplifying assumptions, e.g. omit the losses in pressure on pipelines and heat exchangers as well as, which is the most significant, use cycles (i.e. closed systems; Chap. 2) in real "open" thermodynamic processes analyses occurring in *gas–gas* systems, i.e. processes input and output with thermodynamic media.

To sum up, making the above-mentioned simplifying assumptions perfectly allowed us to perform thermodynamic and economic analyses and draw generalised conclusions from them concerning *gas–gas* systems.

References

1. Bartnik R (2009) Gas-steam power plants and combined heat and power plants. Energy and economic efficiency (in Polish: Elektrownie i elektrociepłownie gazowoparowe. Efektywność energetyczna i ekonomiczna), WNT, Warszawa 2009 (reprint 2012, 2017)
2. Bartnik R, Buryn Z, Hnydiuk-Stefan A, Skomudek W (2020) Power Plant Retrofit and Modernization (in Polish: Modernizacja elektrowni. Efektywność energetyczna i ekonomiczna), WNT, Warszawa
3. Bartnik R, Buryn Z, Hnydiuk-Stefan A, Skomudek W (2019) Dual-fuel gas-steam combined heat and power plants (in polish: dwupaliwowe elektrownie i elektrociepłownie gazowo-parowe. Efektywność energetyczna i ekonomiczna), WNT, Warszawa

4. Bartnik R (2013) The modernization potential of gas turbines in the coal-fired power industry. Thermal and Economic Effectiveness, Springer, London
5. Bartnik R, Buryn Z, Hnydiuk-Stefan A (2017) Methodology and a continuous time mathematical model for selecting the optimum capacity of a heat accumulator integrated with a CHP plant (in Polish: Ekonomika Energetyki w modelach matematycznych z czasem ciągłym). Wydawnictwo Naukowe PWN, Warszawa
6. Bartnik R, Bartnik B (2014) Economic calculation in the energy sector (in Polish: Rachunek ekonomiczny w energetyce). WNT, Warszawa

Chapter 2
Basic Thermodynamic Analyses of Hierarchical Systems

The losses of exergy stream (reducing mechanical power) in the thermodynamic system caused by the increase in entropy for input and output power media taking part in inner thermodynamic processes and increase in entropy of external heat sources contacting them:

$$\delta \dot{B} = T_{amb}\left(\sum_{k} \Delta \dot{S}_{med} + \sum_{l} \Delta \dot{S}_{so}\right) \tag{2.1}$$

where:

k number of media input and output from the system,
l number of external heat sources, including the ambience, contacting the system.

Here it should be noted that the increase in entropy of the external heat source at the temperature T_{so} = const providing the heat stream $\dot{Q}$ to the system can be calculated using the entropy definition:

$$\Delta \dot{S}_{so} = -\int \frac{d\dot{Q}}{T_{so}} = -\frac{\dot{Q}}{T_{so}} \tag{2.2}$$

The minus character in (2.2) means that the positive heat stream $\dot{Q}$ is output from the source. For the source collecting the heat stream $\dot{Q}$ from the system, in (2.2) only the character changes:

$$\Delta \dot{S}_{so} = \int \frac{d\dot{Q}}{T_{so}} = +\frac{\dot{Q}}{T_{so}} \tag{2.3}$$

It should be noted that the Carnot engine:

$$N_C = \eta_C \dot{Q} \tag{2.4}$$

R. Bartnik and T. W. Kowalczyk, *Hierarchical Gas-Gas Systems*,
Power Systems, https://doi.org/10.1007/978-3-030-69205-6_2

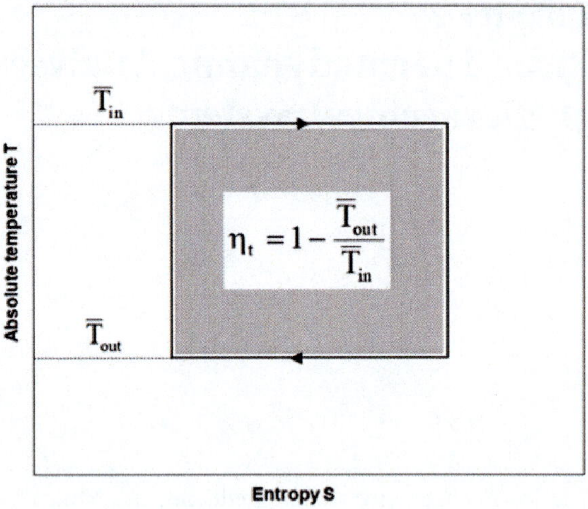

Fig. 2.1 Thermodynamic cycle of the engine

corresponds to exergy stream $\dot{B}_{\dot{Q}}$ heat stream $\dot{Q}$ collected from the source at the temperature of T_g = const, $N_C \equiv \dot{B}_{\dot{Q}}$ (i.e. maximum power output that can be obtained under specific ambient conditions from a specific thermodynamic medium or a specific exergy stream; mechanical work exergy L or power N, therefore it equals directly to work or power, $B_L = L$, $\dot{B}_L = N$). Power $N_C \equiv \dot{B}_{\dot{Q}}$ corresponds to the rectangle area drawn with a broken line in Fig. 1.1. (see Fig. 2.1). The Carnot engine has the maximum potential efficiency for generating electricity in mechanical thermodynamic systems:

$$\eta_C = 1 - \frac{T_{amb}}{T_g} < 1 \tag{2.5}$$

where:

T_{amb} absolute temperature of a lower heat source (ambience),
T_g absolute temperature of an upper heat source.

As we can see in (2.4) and (2.5), which is extremely important, the same heat stream $\dot{Q}$ collected from a hot source with the temperature T_g = const can be converted to mechanical power in higher level, the higher the temperature T_g. The exergy loss caused only by temperature reduction from $T_{g\,1}$ to $T_{g\,2}$ ($T_{g\,1} > T_{g\,2}$) is:

$$\delta \dot{B}_{\dot{Q}} = \Delta N_C = \dot{Q}\left(1 - \frac{T_{amb}}{T_{g\,1}}\right) - \dot{Q}\left(1 - \frac{T_{amb}}{T_{g\,2}}\right) = T_{amb}\dot{Q}\frac{T_{g\,1} - T_{g\,2}}{T_{g\,1}T_{g\,2}} \tag{2.6}$$

This loss can be also interpreted as a loss caused by irreversibility of the heat stream $\dot{Q}$ between two sources with the temperature T_{g1} = const and T_{g2} = const. Therefore, (2.6) can be also obtained by using dependencies (2.1), (2.2) and (2.3):

$$\delta\dot{B}_{\dot{Q}} = T_{amb}\sum_{l=2}\Delta\dot{S}_{so} = T_{amb}\left(-\frac{\dot{Q}}{T_{g1}} + \frac{\dot{Q}}{T_{g2}}\right). \tag{2.7}$$

To analyse real thermodynamic processes that occur in open systems, it is convenient to use comparative thermodynamic cycles (closed systems). Furthermore, if we introduce entropic average temperatures in the cycles for processes with heat absorption or emission by heat transfer media, then the thermodynamic analysis of those processes will be additionally facilitated. For the thermodynamic medium the average thermodynamic temperature during heat absorption with no effective work in isobaric process 1–2 of the *Joule-Brayton* cycle can be calculated from the equation:

$$\overline{T}_{in} = \frac{Q_{in}}{\Delta S} = \frac{\int_{s_1}^{s_2} T_{in}(s)ds}{s_2 - s_1} = \frac{i_2 - i_1}{s_2 - s_1} \tag{2.8}$$

where:

ΔS increase in entropy of the heat transfer medium during isobaric process 1–2 of heat input to it Q_{in},

i, s specific enthalpy and entropy of the heat transfer medium.

The same method of definition applies to temperature $\overline{T}_{out}$ during heat output Q_{out} from the heat transfer medium during an isobaric process.

Replacing real processes with open cycles (closed cycles) causes that there are not heat transfer media input or output from the system, i.e.$\sum_k \Delta S_{med} = 0$ (2.1). Also for the medium circulating in the thermodynamic cycle, the increase in entropy is zero, $\sum \Delta S_{med}^{cycle} = 0$, because its final state is identical with the initial state. Then, the increase in entropy for all bodies taking part in the phenomenon expresses only with the increase in the entropy of heat sources that can be expressed by the sum of entropy increase at an irreversible heat flow between sources and media and between media during absorption and emission of heat (see (2.15)). Also the general form of system exergy balance:

$$\dot{B}_{in} = \dot{B}_{out} + N + \sum_m \Delta\dot{B}_{so} + \delta\dot{B} \tag{2.9}$$

then reduces to dependency:

$$0 = N + \sum_m \Delta\dot{B}_{so} + \delta\dot{B} \tag{2.10}$$

while the increase in the exergy of external source being in contact with the system can be expressed by the following formula (refer to (2.4), (2.5)):

$$\Delta\dot{B}_{so} = \pm\,\dot{Q}_{so}\frac{T_{so} - T_{amb}}{T_{so}} \tag{2.11}$$

where:

$\dot{B}_{in}, \dot{B}_{out}$	exergy streams of media input and output from the system,
$\sum_m \Delta\dot{B}_{so}$	sum of increases in the exergy of the external heat source (without ambience) being in contact with the system (ambience as per definition is an energetically worthless heat source),
m	number of external heat sources (without ambience) contacting the system. ($m = l - 1$, refer to (2.1)),
N	power output from the system,
$\dot{Q}_{so}$	heat stream input or output from the heat source,
T_{so}	heat source absolute temperature,
$\delta\dot{B}$	internal losses in exergy caused by irreversibility of heat flow $\delta\dot{B} = T_{amb}\sum_l \Delta\dot{S}_{so}$(2.1).

The minus sign in (2.11) corresponds to the situation when the heat stream $\dot{Q}_{so}$ is returned to the system by the source. However, this is obvious that when the temperature of the heat source is lower than ambient temperature, $T_{so} < T_{amb}$, its exergy increases when emitting heat from it, and conversely, when source exergy decreases $T_{so} > T_{amb}$, because its capacity to provide work is reduced. By contrast, the plus sign in (2.11) corresponds to the situation when the heat stream $\dot{Q}_{so}$ is input from the system to the external source. It is also obvious that ambience ($T_{so} = T_{amb}$) constitutes a source of completely worthless heat in terms of power generation purposes, therefore its exergy, according to the definition, is zero and does not change, no matter if the heat is input or output.

2.1 Entropic Average Temperatures

Introducing average thermodynamic temperatures $\overline{T}_{in}$ and $\overline{T}_{out}$ (2.8) respectively for heat input and output from the cycle (determined for real temperatures and pressures that occur at the beginning and end of those processes, i.e. for irreversible processes), allows us to give a rectangular shape to any cycle in *temperature-entropy* coordinate system (Fig. 2.1), no matter what the nature of remaining physical processes inside it, the ones with effective work, whether they are irreversible or not. By analogy to (2.5) the power generation efficiency of any cycle can be expressed with the following formula:

$$\eta_t = 1 - \frac{\overline{T}_{out}}{\overline{T}_{in}} \tag{2.12}$$

where the temperature of T_{amb}Carnot cycle isotherm of the lower heat source, i.e. ambience, is replaced by entropic average temperature $\overline{T}_{out}$ during outputting heat from any cycle, while temperature T_g Carnot cycle isotherm equalling the upper heat source temperature is replaced by entropic average temperature $\overline{T}_{in}$ during inputting heat to the medium in any cycle under consideration—Fig. 2.1 Dependency (2.12) shows that generating electricity in engine cycles should take place at the temperature $\overline{T}_{in}$ of the heat transfer medium as high as possible during collecting power heat, i.e. heat from an external heat source, and at a temperature $\overline{T}_{out}$ of the heat transfer medium as low as possible during bringing heat back to the cycle.

By contrast, the loss in exergy caused by friction in the above-mentioned cycle processes, where effective work takes place, i.e. when they are irreversible, which obviously takes place, it should only be properly adjusted to the assumed correction mode in entropic average temperatures of inputting $\overline{T}_{in}$ and outputting heat $\overline{T}_{out}$ from the cycle. The ratio of corrected average temperatures must obviously equal the ratio of heat output Q_{out} and input Q_{in} to a specific cycle (2.13). For instance, in an irreversible *Joule-Brayton* cycle, being a comparative cycle for a gas turbine engine (Fig. 2.2), for which, as for every engine no matter what friction, the following power balance equation applies:

$$N = \dot{Q}_{in} - \dot{Q}_{out} = \dot{Q}_{in}\left(1 - \frac{\dot{Q}_{out}}{\dot{Q}_{in}}\right) = \dot{Q}_{in}\left(1 - \frac{\bar{T}_{out\ s}\Delta\dot{S}_{out}}{\bar{T}_{in\ s}\Delta\dot{S}_{in}}\right) = \dot{Q}_{in}\left(1 - \frac{\bar{T}_{out\ s}}{\bar{T}^{cor}_{in}}\right). \quad (2.13)$$

The best solution would be to reduce only the average thermodynamic temperature during absorption by a heat medium adding to the entropy increase, taking place in this process, the increases of entropy in irreversible adiabatic processes of its compression and expansion.

The corrected average temperature during heat absorption would then equal the ratio of entropic average temperature for the process and the ratio of entropy increases

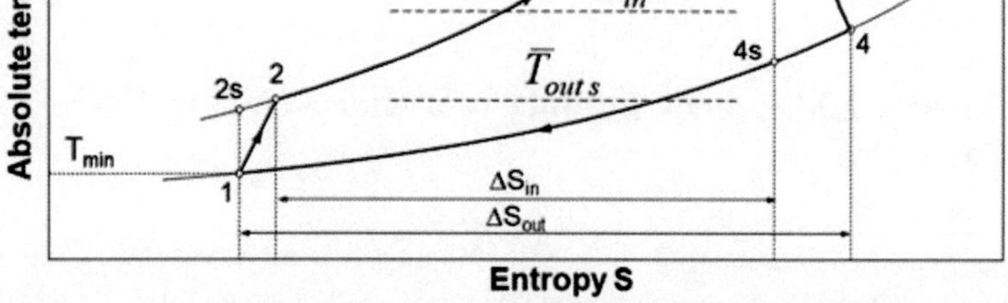

Fig. 2.2 The processes of the thermodynamic medium in a comparative Joule-Brayton cycle of a gas turbine engine

for heat absorption and emission (2.14). The average thermodynamic temperature during heat emission would not change and would equal average actual temperature.

The corrected (reduced) temperature during heat absorption Q_{in} by heat transfer medium would be expressed by the following formula:

$$\overline{T}_{in}^{cor} = \overline{T}_{in\ s} \frac{\Delta S_{in}}{\Delta S_{in} + \Delta S_{com} + \Delta S_{ex}} = \overline{T}_{in\ s} \frac{\Delta S_{in}}{\Delta S_{out}} \tag{2.14}$$

where in the above presented formulas:

ΔS_{ex}, ΔS_{com} increases of medium entropy in irreversible adiabatic processes of expansion and compression of the heat transfer medium,

ΔS_{in} increase in heat transfer medium entropy in the process of heat absorption,

ΔS_{out} increase in heat transfer medium entropy in the process of heat emission,

$\overline{T}_{in\ s}$ entropic average temperature during the isobaric heat absorption process Q_{in} by heat transfer medium (when the isobar is subjected to friction, then determining by using (2.8) the temperature $\overline{T}_{in\ s}$, it is necessary to consider the increase in entropy of the medium caused by pressure).

$\overline{T}_{out\ s}$ entropic average temperature during the isobaric heat emission process Q_{out} by heat transfer medium (when the isobar is subjected to friction, then determining by using (2.8) the temperature $\overline{T}_{out\ s}$, it is necessary to consider the increase in entropy of the medium caused by pressure).

When determining entropic average temperatures (2.8), it can additionally include different medium stream masses in individual places of heat absorption and emission processes as it is the case in real open systems. This is the way to include the increase in entropy of media, which takes place in real processes, i.e. when power media are input or output to the heat system, which corresponds to the following situation $\sum_k \Delta S_{med} \neq 0$ [1].

2.2 Energy and Exergy Balance of a Hierarchical *j*-Cycle Engine

In a general case the number of heat transfer media can be arbitrarily large. Figure 2.3 presents the schematic diagram of a hierarchical *j*-cycle heat engine.

Increasing the number of media with different ranges of operating temperature makes it possible to use a wider temperature range between the temperature of the upper and lower heat source (ambience). Thus, it allows us to reduce the losses of exergy in the system. i.e. increase the electricity production. However, this solution

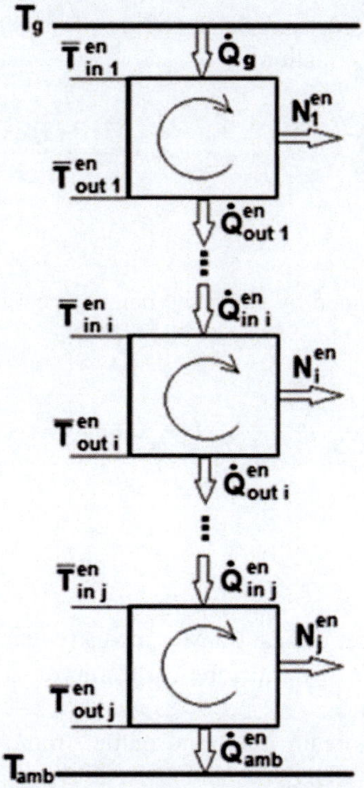

Fig. 2.3 The schematic diagram of a hierarchical j-cycle heat engine; j—number of cycles (engines), N_i^{en}—power i-of the engine ($i = 1 \div j$), $\dot{Q}^{en}_{in\ i}$, $\dot{Q}^{en}_{out\ i}$—input and output heat stream from i-cycle (engine), $\dot{Q}_g \equiv \dot{E}_{ch}$—heat stream input to the system from the upper heat source (fuel chemical energy stream), $\dot{Q}^{en}_{amb}$—heat stream emitted to ambience, T_{amb}—absolute ambient temperature (temperature of the lower heat source), T_g—absolute fuel combustion temperature (temperature of the upper heat source), $\overline{T}^{en}_{in\ i}$, $\overline{T}^{en}_{out\ i}$—entropic average temperatures of heat absorbing and emitting medium in *i*—cycle (engine)

has a disadvantage, because it needs higher investment outlays to implement the system. As already mentioned, the loss of exergy in the "*j*-cycle" engine caused by the increase in entropy of heat sources can be divided into the losses caused by the heat flow irreversibility between sources and cycles.

The exergy loss in a closed system with two heat sources ($l = 2$) can be then expressed with dependency (see (2.2), (2.3), (2.6)):

$$\delta \dot{B}_{en} = T_{amb} \sum_{l=2} \Delta \dot{S}_{so} = T_{amb} \left(\frac{\dot{Q}^{en}_{amb}}{T_{amb}} - \frac{\dot{Q}_g}{T_g} \right) = \sum_{i=1}^{j+1} \delta \dot{B}^{en}_i = T_{amb} \sum_{i=1}^{j+1} \dot{Q}^{en}_{in\,i} \frac{\overline{T}^{en}_{out\ i-1} - \overline{T}^{en}_{in\ i}}{\overline{T}^{en}_{out\ i-1} \overline{T}^{en}_{in\ i}} \tag{2.15}$$

while the system power is determined by using the balance exergy (2.10) and is presented in the following equation:

$$N_{en} = N_C^{en} - \delta \dot{B}_{en} = -\sum_{m=1} \Delta \dot{B}_{so} - \delta \dot{B}_{en} = \dot{Q}\frac{T_g - T_{amb}}{T_g} - T_{amb}\left(\frac{\dot{Q}_{amb}^{en}}{T_{amb}} - \frac{\dot{Q}_g}{T_g}\right) \tag{2.16}$$

while the power is determined by using the balance energy (2.13) in the following equation:

$$N_{en} = \dot{Q}_g - \dot{Q}_{amb}^{en} = \sum_{i=1}^{j} N_i^{en} = \sum_{i=1}^{j} \left(\dot{Q}_{in\,i}^{en} - \dot{Q}_{out\,i}^{en}\right) = \sum_{i=1}^{j} \dot{Q}_{in\,i}^{en} \frac{\overline{T}_{in\,i}^{en} - \overline{T}_{out\,i}^{en}}{\overline{T}_{in\,i}^{en}} \tag{2.17}$$

where:

j	number of heat transfer media (engines),
N_C^{en}, N_i^{en}	power of the theoretical Carnot engine and the power of real engines,
$\dot{Q}_{in\,i}^{en}$, $\dot{Q}_{out\,i}^{en}$	heat stream input and output from i-cycle (engine), wherein $\dot{Q}_{out\,i}^{en} = \dot{Q}_{in\,i+1}^{en}$ and $\dot{Q}_{in\,1}^{en} \equiv \dot{Q}_g$, $\dot{Q}_{in\,j+1}^{en} \equiv \dot{Q}_{amb}^{en}$
$\dot{Q}_{amb}^{en}$, $\dot{Q}_g \equiv \dot{E}_{ch}$	heat stream output from the system to ambience and input to the system from the upper heat source (fuel chemical energy stream),
$\overline{T}_{in\,i}^{en}$, $\overline{T}_{out\,i}^{en}$	average thermodynamic temperature of heat absorption and emission medium in i–cycle (engine), wherein $\overline{T}_{in\,j+1}^{en} \equiv T_{amb}$, $\overline{T}_{out\,0}^{en} \equiv T_g$
T_g	absolute temperature of the upper heat source (fuel combustion temperature).

The quantity $\left(\overline{T}_{in\,i}^{en} - \overline{T}_{out\,i}^{en}\right)/\overline{T}_{in\,i}^{en}$ in the last formula part (2.17) expresses power efficiency i-cycle ($i = 1 \div j$) engine operating between entropic average temperatures in real processes of inputting and outputting heat $\overline{T}_{in\,i}^{en}$, $\overline{T}_{out\,i}^{en}$ (refer to (2.12), Fig. 2.1):

$$\eta_i = \frac{N_i^{en}}{\dot{Q}_{in\,i}^{en}} = 1 - \frac{\overline{T}_{out\,i}^{en}}{\overline{T}_{in\,i}^{en}} \tag{2.18}$$

The heat stream output from i-cycle (engine) using the entropic average temperatures can be only expressed by heat stream $\dot{Q}_{in\,1}^{en} \equiv \dot{Q}_g \equiv \dot{E}_{ch}$ delivered to the system from a source with the temperature $\overline{T}_{out\,0}^{en} \equiv T_g$.

$$\dot{Q}^{en}_{out\ i} = \dot{Q}^{en}_{in\ i}\frac{\overline{T}^{en}_{out\ i}}{\overline{T}^{en}_{in\ i}} = \dot{Q}^{en}_{in\ i+1} = \dot{Q}^{en}_{in\ 1}\prod_{n=1}^{i}\frac{\overline{T}^{en}_{out\ n}}{\overline{T}^{en}_{in\ n}} \tag{2.19}$$

Formula (2.19) converts into a dependency to express the heating heat stream output from the system to ambience.

$$Q^{en}_{amb} = Q^{en}_{in\,1}\prod_{i=1}^{j}\frac{\bar{T}^{en}_{out\,i}}{\bar{T}^{en}_{in\,i}} \tag{2.20}$$

Using additionally dependency (2.19) formula (2.17) for the total system power can adopt the following form.

$$N_{en} = \dot{Q}_g - \dot{Q}^{en}_{amb} = \sum_{i=1}^{j} N^{en}_i = \sum_{i=1}^{j}(\dot{Q}^{en}_{in\ i} - \dot{Q}^{en}_{out\ i}) = \dot{Q}^{en}_{in\ 1}\left(1 - \prod_{i=1}^{j}\frac{\overline{T}^{en}_{out\ i}}{\overline{T}^{en}_{in\ i}}\right) \tag{2.21}$$

The quantity in brackets on the right-hand side of the formula (2.21) means power efficiency η_{1-j} for producing electricity in *j*-cycle system expressed by using entropic average temperatures.

$$\eta_{1-j} = 1 - \prod_{i=1}^{j}\frac{\overline{T}^{en}_{out\ i}}{\overline{T}^{en}_{in\ i}} \tag{2.22}$$

For example, for *2*-cycle system the efficiency, using additionally dependency (2.18), can be expressed by the following formula.

$$\eta_{1-2} = 1 - \frac{\bar{T}^{en}_{out\,1}}{\bar{T}^{en}_{in\,1}} - \frac{\bar{T}^{en}_{out\,2}}{\bar{T}^{en}_{in\,2}} = \eta_1 + \eta_2 - \eta_1\eta_2. \tag{2.23}$$

The final form of (2.15) that "distinguishes" the places of loss in exergy in the system allows us to analyse the ways of its thermodynamic improvements. This is the case, because it indicates the places of the biggest exergy losses that determine its low exergetic efficiency, thus indicating the places where it should be improved. It indicates entropic average temperatures and necessary conditions for changing its values to keep improving the system thermodynamic perfection.

In addition, it allows us to easily assess the quantity of reasons that reduce the perfection. Furthermore, it allows us to quickly analyse the changes in parameters for preceding cycles resulting in changes in exergy losses in succeeding cycles, thus resulting in changes in exergy losses in the whole system. Therefore, it allows us to evaluate the appropriateness of those changes not only at the design stage but also in case of modernization [2].

The final form of formula (2.15) makes it possible to analyse the impact of power efficiency of individual engines (entropic average media temperatures) on the total exergetic efficiency of the system.

As already indicated, the irreversible heat flow is the biggest source of losses in exergy of the system (excluding the exergy loss during combustion). The losses in exergy resulting in performing the processes in real conditions, losses of hydraulic friction, losses of mixing substances with unequal temperatures etc. are as compared to them in the real processes of electricity and heat generation fairly frequently low, thus, when their value is significantly low, they can be neglected.

In the method of entropic average temperatures the losses, as already mentioned, can be considered in the thermodynamic temperatures determined for the processes of heat absorption and emission by the heat transfer medium adjusting it by considering the increases in entropy occurring in remaining processes (2.14).

A detailed thermodynamic and economic analysis of a hierarchical *gas–gas* engine is presented in Chap. 3.

2.3 Energy and Exergy Balance of a Hierarchical *j*-Cycle of a Compressor Chiller

Figure 2.4 presents a diagram of a tricycle ($j = 3$, refer to formulas (2.24)–(2.31)compressor chiller in the hierarchical (cascade) system. Combining subsequent cycles (stages) of a chiller, lower in the hierarchy just above them, takes place in heat exchangers operating at the same time as condensers for lower cycles and evaporators for higher cycles.

The exergy loss generally in the *j*-cycle system can be expressed in the following way:

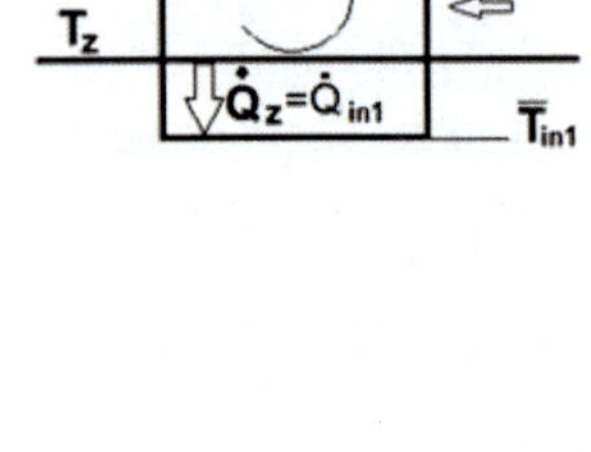

Fig. 2.4 The schematic diagram of a tricycle compressor chiller

$$\delta \dot{B}_z = T_{amb} \sum_{l=2} \Delta \dot{S}_{so} = T_{amb}\left(\frac{\dot{Q}_{amb}}{T_{amb}} - \frac{\dot{Q}_z}{T_z}\right) = \sum_{i=1}^{j+1} \delta \dot{B}_{zi} = T_{amb} \sum_{i=1}^{j+1} \dot{Q}_{in\ i} \frac{\overline{T}_{out\ i-1} - \overline{T}_{in\ i}}{\overline{T}_{out\ i-1} \overline{T}_{in\ i}} \tag{2.24}$$

While the system driving power determined by exergy balance can be expressed with equation (refer to (2.16)):

$$N_z = N_{C\,z} + \delta \dot{B}_z = \sum_{m=1} \Delta \dot{B}_{so} + \delta \dot{B}_z = -\dot{Q}_z \frac{T_z - T_{amb}}{T_z} + T_{amb}\left(\frac{\dot{Q}_{amb}}{T_{amb}} - \frac{\dot{Q}_z}{T_z}\right) \tag{2.25}$$

while the power is determined by using the balance energy in the following formula:

$$N_z = \dot{Q}_{amb} - \dot{Q}_z = \sum_{i=1}^{j} N_i = \sum_{i=1}^{j} (\dot{Q}_{out\ i} - \dot{Q}_{in\ i}) = \sum_{i=1}^{j} \dot{Q}_{in\ i} \frac{\bar{T}_{out\ i} - \bar{T}_{in\ i}}{\bar{T}_{in\ i}} \tag{2.26}$$

where:

j	number of heat transfer media (compressor chillers),
$N_{C\,z}$, N_i	driving power of a theoretical chiller operating in the reversible anticlockwise Carnot cycle and the capacities of real compressor chillers,
$\dot{Q}_{in\ i}$, $\dot{Q}_{out\ i}$	input and output heat stream from *i*-cycle of the chiller, wherein $\dot{Q}_{out\ i} = \dot{Q}_{in\ i+1}$ and $\dot{Q}_{in\ 1} \equiv \dot{Q}_z$, $\dot{Q}_{in\ j+1} \equiv \dot{Q}_{amb}$
$\dot{Q}_{amb}$, $\dot{Q}_z$	heat streams output from the system to ambience and input to the system from the lower heat source (chiller chamber),
$\overline{T}_{in\ i}$, $\overline{T}_{out\ i}$	average thermodynamic temperature of heat absorption and emission medium in *i*–cycle of the compressor chiller, wherein $\overline{T}_{in\ j+1} \equiv T_{amb}$, $\overline{T}_{out\ 0} \equiv T_z$
T_z	absolute temperature of the lower heat source (temperature in the chiller chamber).

The inverse quantity $(\overline{T}_{out\ i} - \overline{T}_{in\ i})/\overline{T}_{in\ i}$ in the last formula part (2.26) expresses power efficiency *i*-cycle ($i = 1 \div j$) of a chiller operating between entropic average temperatures in real processes of inputting and outputting heat $\overline{T}_{in\ i}$, $\overline{T}_{out\ i}$.

$$\varepsilon_i = \frac{\dot{Q}_{in\ i}}{N_i} = \frac{\dot{Q}_{in\ i}}{\dot{Q}_{out\ i} - \dot{Q}_{in\ i}} = \frac{\bar{T}_{in\ i}}{\bar{T}_{out\ i} - \bar{T}_{in\ i}} \tag{2.27}$$

Efficiency ε_i can be higher or lower than one.

The heat stream output from *i*-cycle of the chiller using the entropic average temperatures can be only expressed by heat stream $\dot{Q}_{z\ in\ 1} \equiv \dot{Q}_z$ delivered to the system from a source at the temperature $\overline{T}_{z\ out\ 0} \equiv T_z$.

$$\dot{Q}_{out\ i} = \dot{Q}_{in\ i}\frac{\overline{T}_{out\ i}}{\overline{T}_{in\ i}} = \dot{Q}_{in\ i+1} = \dot{Q}_z\prod_{n=1}^{i}\frac{\overline{T}_{out\ n}}{\overline{T}_{in\ n}} \tag{2.28}$$

From formula (2.28) we receive the dependency for the heat stream output to ambience from a hierarchical chiller system.

$$\dot{Q}^{1-j}_{amb} = \dot{Q}_z\prod_{i=1}^{j}\frac{\overline{T}_{out\ i}}{\overline{T}_{in\ i}} \tag{2.29}$$

Using additionally dependency (2.28) formula (2.26) for the total system power can adopt the following dependency.

$$N_z = \sum_{i=1}^{j} N_i = \sum_{i=1}^{j}(\dot{Q}_{out\ i} - \dot{Q}_{in\ i}) = \dot{Q}_z\left(\prod_{i=1}^{j}\frac{\bar{T}_{out\ i}}{\bar{T}_{in\ i}} - 1\right) \tag{2.30}$$

The quantity in brackets on the right-hand side of the (2.30) means inverse power efficiency ε_{1-j} for producing chill in the *j*-cycle of the hierarchical chiller system expressed by using entropic average temperatures.

$$\varepsilon_{1-j} = \frac{\dot{Q}_z}{N_z} = \frac{1}{\prod_{i=1}^{j}\frac{\overline{T}_{out\ i}}{\overline{T}_{in\ i}} - 1} \tag{2.31}$$

For example, for 2-cycle system the efficiency can be expressed by the following formula.

$$\varepsilon_{1-2} = \frac{1}{\frac{\bar{T}_{out\ 1}}{\bar{T}_{in\ 1}}\frac{\bar{T}_{out\ 2}}{\bar{T}_{in\ 2}} - 1} = \frac{\varepsilon_1\varepsilon_2}{\varepsilon_1 + \varepsilon_2 + 1} \tag{2.32}$$

Formula (2.29) for a single-cycle system, $j = 1$, heat $\dot{Q}^1_{amb}$ output from it to the ambience is lower than the heat $\dot{Q}^{1-j}_{amb}$ output from *j-cycle* at identical extreme temperatures: $\overline{T}^1_{in\ 1} = \overline{T}^{1-j}_{in\ 1}, \overline{T}^1_{out\ 1} = \overline{T}^{1-j}_{out\ j}$ and identical value of the heat stream output from the chiller chamber $\dot{Q}_z$.

$$\dot{Q}^1_{amb} < \dot{Q}^{1-j}_{amb} \tag{2.33}$$

Thus, higher are the losses of exergy in the hierarchical system (2.24) and higher, as compared to a single-cycle chiller, its driving power (2.25), (2.26). Furthermore, the higher the number of cycles, the higher the losses of exergy and power. Increasing their number in the chiller (generally speaking in a working machine) is, contrary to the hierarchical engine, unfavourable. Also, the increase in the number of cycles raises, obviously, investment outlays on the chiller. The number of chiller cycles should be as low as possible. The construction of the hierarchical chiller makes sense only when the temperature in the chiller chamber T_z is so low that it is impossible to choose a heat transfer medium showing desired properties within the whole $\langle T_z;\, T_{amb}\rangle$ temperature range and that is why it is necessary to use hierarchical chillers where at least two chiller cycles work together.

2.4 Thermodynamic Analysis of a Hierarchical Gas–Gas Engine System Driving a Compressor Chiller

Figure 2.5 presents the hierarchical *gas–gas* engine system driving the compressor chiller.

The balance of system exergy (engine + chiller) for its steady state operation is expressed by dependency (2.10):

$$0 = \sum_{m=2} \Delta \dot{B}_{so} + N_{en} - N_z + \delta \dot{B} \tag{2.34}$$

where:

$\Delta \dot{B}_{so}$ increase in exergy for the external source delivering energy to the system,

N_{en} power of hierarchical gas–gas engine; engine power is the sum of power for gas turbine and turboexpander, $N_{en} = N^{GT} + N^{TE}$,

N_z chiller driving power.

The sum of the increases of external exergy sources delivering heat to the system can be expressed by the following dependency (refer to (2.11)).

$$\sum_{m=2} \Delta \dot{B}_{so} = \Delta \dot{B}_{so}^{T_g} + \Delta \dot{B}_{so}^{T_z} = -\dot{Q}_g \frac{T_g - T_{amb}}{T_g} - \dot{Q}_z \frac{T_z - T_{amb}}{T_z} \tag{2.35}$$

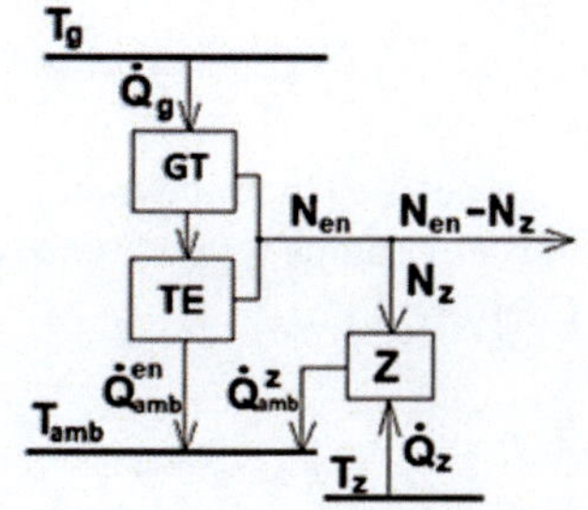

Fig. 2.5 The schematic diagram of a hierarchical gas–gas engine (gas turbine GT plus turboexpander TE) driving the compressor chiller Z

Formula (2.35) shows that the exergy of the heat source with the temperature T_g, from which a driving heat stream is collected $\dot{Q}_g$ decreases, wherein the loss in exergy $\Delta \dot{B}_{so}^{T_g}$ is lower than the amount of heat $\dot{Q}_g$. In the case of the chiller, the exergy of the chiller chamber with the temperature T_z, from which the heat stream is collected $\dot{Q}_z$ increases, wherein the directions of heat $\dot{Q}_z$ and exergy streams $\Delta \dot{B}_{so}^{T_z}$ are, in this case, opposite.

The exergy loss in the system can be expressed by the following dependency:

$$\delta \dot{B} = T_{amb} \sum_{l=3} \Delta \dot{S}_{so} = T_{amb} \left(\frac{\dot{Q}_{amb}^{en} + \dot{Q}_{amb}^{z}}{T_{amb}} - \frac{\dot{Q}_g}{T_g} - \frac{\dot{Q}_z}{T_z} \right) \quad (2.36)$$

wherein the heat stream $\dot{Q}_g$ corresponds to the fuel chemical energy stream burnt in the gas turbine combustion chamber, $\dot{Q}_g \equiv \dot{E}_{ch}$. The temperature of fuel combustion is T_g.

The power output from the system can be determined, for instance, using the formula of exergy balance (2.34) substituting there Eqs. (2.35) and (2.36) (compare formulas (2.16), (2.25)):

$$N_{en} - N_z = \dot{Q}_g \frac{T_g - T_{amb}}{T_g} + \dot{Q}_z \frac{T_z - T_{amb}}{T_z} - \delta \dot{B} = \dot{Q}_g + \dot{Q}_z - \dot{Q}_{amb}^{en} - \dot{Q}_{amb}^{z} \quad (2.37)$$

Equation (2.37), its final form, can be obviously obtained directly from the balance of system energy. It is presented both by its left and right-hand side of the equation.

Heat streams output to the ambience from the hierarchical double-cycle engine $\dot{Q}_{amb}^{en}$ and the most advantageous in terms of thermodynamics and economy single-cycle chiller $\dot{Q}_{amb}^{z}$ can be expressed by the following equations (refer to (2.20), (2.29)):

$$\begin{aligned} \dot{Q}_{amb}^{en} &= \dot{Q}_g \prod_{i=1}^{2} \frac{\bar{T}_{out\,i}}{\bar{T}_{in\,i}} = \dot{Q}_g \frac{\bar{T}_{out\,1}}{\bar{T}_{in\,1}} \frac{\bar{T}_{out\,2}}{\bar{T}_{in\,2}} = \dot{Q}_g (1 - \eta_{1-2}) \\ &= \dot{Q}_g [1 - (\eta_{TG} + \eta_{TE} - \eta_{TG} \eta_{TE})] \end{aligned} \quad (2.38)$$

$$\dot{Q}_{amb}^{z} = \dot{Q}_z \frac{\bar{T}_{out\ 1}}{\bar{T}_{in\ 1}} = \dot{Q}_z \left(\frac{1}{\varepsilon} + 1 \right) \quad (2.39)$$

A detailed thermodynamic and economic trigeneration system with a hierarchical *gas–gas* engine for combined electricity, heat and cooling generation is presented in Chap. 4.

2.5 Thermodynamic Analysis of a Hierarchical Gas–Gas Engine System Driving a Compressor Heat Pump

Figure 2.6 presents the system of the hierarchical *gas–gas* engine driving the compressor heat pump that delivers heating heat to a room at the temperature T_p.

The sum of exergy of external heat sources contacting the system is expressed by dependency (refer to formula (2.11)); the ambience exergy is zero, because according to the definition, it is the source of worthless heat in terms of power generation purposes.

$$\sum_{m=2} \Delta \dot{B}_{so} = \Delta \dot{B}_{so}^{T_g} + \Delta \dot{B}_{so}^{T_p} = -\dot{Q}_g \frac{T_g - T_{amb}}{T_g} + \dot{Q}^{HP} \frac{T_p - T_{amb}}{T_p} \tag{2.40}$$

The exergy loss in the system can be expressed by the following dependency:

$$\delta \dot{B} = T_{amb} \sum_{l=3} \Delta \dot{S}_{so} = T_{amb} \left(\frac{\dot{Q}_{amb}^{en} - \dot{Q}_{amb}^{HP}}{T_{amb}} - \frac{\dot{Q}_g}{T_g} + \frac{\dot{Q}^{HP}}{T_p} \right) \tag{2.41}$$

The power output from the system can be determined, for instance, by using the formula of exergy balance (2.34) substituting there Eqs. (2.40) and (2.41) (compare formulas (2.16), (2.25)):

$$\begin{aligned} N_{en} - N_{HP} &= \dot{Q}_g \frac{T_g - T_{amb}}{T_g} - \dot{Q}^{HP} \frac{T_p - T_{amb}}{T_p} - T_{amb} \left(\frac{\dot{Q}_{amb}^{en} - \dot{Q}_{amb}^{HP}}{T_{amb}} - \frac{\dot{Q}_g}{T_g} + \frac{\dot{Q}^{HP}}{T_p} \right) = \\ &= \dot{Q}_g - \dot{Q}^{HP} + \dot{Q}_{amb}^{HP} - \dot{Q}_{amb}^{en} \end{aligned} \tag{2.42}$$

In the above formulas $\dot{Q}_g$ corresponds to the fuel chemical energy burnt in the gas turbine combustion chamber, $\dot{Q}_g \equiv \dot{E}_{ch}$. The temperature of fuel combustion is T_g. The left and right-hand side of the Eq. (2.42) present obviously the balance of system energy.

Figure 2.7 presents the strip-based energy balance of the heat pump powered by electric energy generated in any source with a generation efficiency η_{el}, i.e. also in a

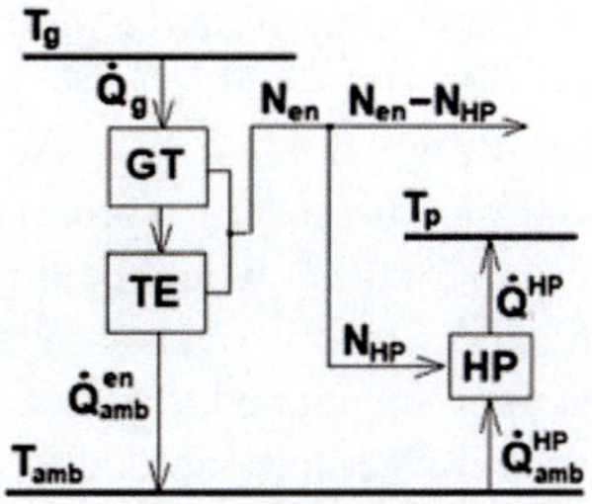

Fig. 2.6 The schematic diagram of a hierarchical gas–gas engine (gas turbine GT plus turboexpander TE) driving the compressor heat pump HP

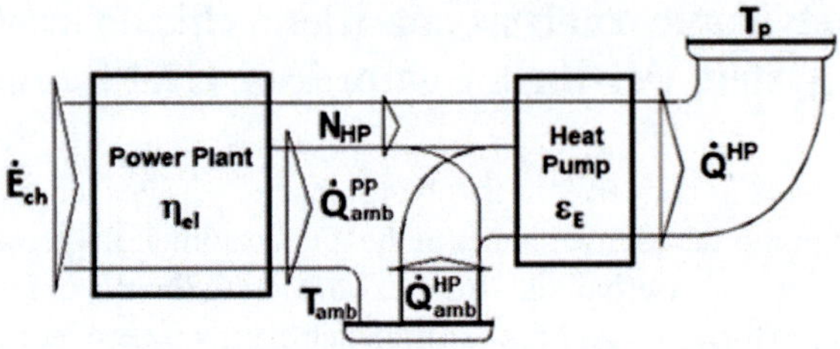

Fig. 2.7 Stripchart of compressor heat pump energy balance: N_{PC}—payable electricity used to drive the pump, $\dot{Q}^{HP}$—heating heat stream delivered to the heating space, $\dot{Q}^{HP}_{amb}$—free heat stream acquired from the ambience, $\dot{Q}^{PP}_{amb}$—heat stream output to the ambience from the power plant producing electricity to drive the pump, ε_E—heat pump energy efficiency index, η_{el} —efficiency of the electricity source

conventional power plant. This balance allows us to better understand the in-depth analysis presented below that answers the question: whether the use of a compressor heat pump is actually thermodynamically viable? By preceding the results of this analysis we can say that it is not, the heat pump is thermodynamically unviable. Furthermore, it is even more unviable in economic terms.

It is obvious that the values from Fig. 2.7 are subject to the following interdependencies:

$$\dot{Q}^{HP} = \dot{Q}^{HP}_{amb} + N_{HP} \tag{2.43}$$

$$\eta_{el} = \frac{N_{HP}}{\dot{E}_{ch}} \tag{2.44}$$

$$\dot{Q}^{PP}_{amb} = \dot{E}_{ch}(1 - \eta_{el}) \tag{2.45}$$

and the so-called heat pump energy efficiency index is expressed by the following formula.

$$\varepsilon_E = \frac{\dot{Q}^{HP}}{N_{HP}} > 1 \tag{2.46}$$

In practice, the index ε_E assumes a low value, just $\varepsilon_E \approx 2,5$.

In the case of delivering heat to a single-family house with a power demand of $\dot{Q}^{HP} = 10$ kW, the electric power needed to drive the heat pump compressor would be as high as $N_{HP} = 4$ kW. The pump would be then a big electricity consumer and thus its cost at a price of 0.6 PLN/kWh and at an annual pump operation of $\tau_A = 2500$ h/a would be high and amounting to as much as PLN 6,000. Then the higher, which is obvious, the value ε_E, the cost would be lower. Unfortunately, as already mentioned, the real value ε_E is low and is just $\varepsilon_E \approx 2,5$. As high as the cost of electricity needed to drive the pump would be the annual capital cost of its operation (the sum of depreciation and financial costs, i.e. costs to return investment outlays incurred for pump "turnkey condition" including interests [1, 3, 4]). Capital costs should be complimented with the annual cost of its maintenance and repairs that depends on those outlays. As standard, it should be assumed that it amounts to approx. 3% of investment outlays. A high amount of the capital cost as well as maintenance and repair cost results from high investment outlay on the pump (the term pump has here more general meaning and includes the very pump and all necessary devices such as heaters, connections infrastructure, automation, etc.). A specially high share of investment outlays have financial resources for installation-construction work, especially for installing the pump evaporator in the soil at a depth of approx. 1.5 m on the surface area of at least 500 m^2 for a single-family house under consideration. Assuming investment outlays on the whole "turnkey" heating system including the pump with a thermal power $\dot{Q}^{HP} = 10$ kW amounting to only PLN 50,000, the annual capital cost including the cost of maintenance and repairs, assuming the depreciation period of 20 years, would be approx. PLN 5,700 (capital cost PLN 4,200, the cost of maintenance and repairs PLN 1,500). Then the annual cost of heat pump operation would be as high as PLN 11,700, which means that its use is completely economically unviable. To make the use of the pump more economically viable than a gas boiler room, the value of its energy efficiency index would have to exceed $\varepsilon_E > 9$ [4], which cannot be obtained under real operating conditions. Even higher would have to be index ε_E, when the pump should replace a coal boiler room, because the coal used to fire it is more than twice cheaper per unit of chemical fuel energy than gas. However, sometimes the decision to install the pump may be made for non-economic reasons, e.g. to improve life comfort, but, as already mentioned, it will be costly. So, in conclusion, we can (should) formulate a general truth that to achieve something contrary to natural processes, e.g. to force the heat flow from a lower to higher temperature, we should take into account high costs.

A thorough pump comparative analysis, both thermodynamic and economic with other heat generation sources, including combined heat and power plants according to all available power generation technologies, are presented in the monograph [4].

Going back to the thermodynamic analysis of a compressor heat pump, using (2.44) and assuming the efficiency value $\eta_{el} = 32\%$ (this is an average efficiency value for domestic power plants), then the fuel chemical energy stream burnt in the power plant for $\dot{Q}^{HP} = 10$ kW and $N_{HP} = 4$ kW is $\dot{E}_{ch} = 12,5$ kW. Formula (2.45) shows that the heat stream output from the power plant to the atmosphere is $\dot{Q}^{PP}_{amb} = 8,5$ kW and is by more than ten percent higher than the heat stream $\dot{Q}^{HP}_{amb} = 6$ kW (2.43) taken by the pump from the atmosphere. Therefore, the use of the pump is thermodynamically unviable.

From the condition of thermodynamic "neutrality", i.e. for the situation when the heat stream $\dot{Q}^{HP}_{amb}$ is not lower than the stream $\dot{Q}^{PP}_{amb}$:

$$\dot{Q}^{HP}_{amb} \geq \dot{Q}^{PP}_{amb} \tag{2.47}$$

using formulas (2.43)–(2.46) we obtain a condition that "binds" pump energy efficiency ε_E with efficiency η_{el}:

$$\varepsilon_E \geq \frac{1}{\eta_{el}} \tag{2.48}$$

According to dependency (2.48), the higher the value ε_E, the lower can be the efficiency of power generation in the power plant η_{el}, to make an independently operating heat pump in comparison to it (generally speaking with a source that generates only electricity) thermodynamically viable. For real value $\varepsilon_E \approx 2,5$ the efficiency should be higher $\eta_{el} \geq 0,4$. It is even possible for the gas-gas turbine set—Fig. 3.2. Its efficiency η^{G-G} reaches higher values than 0.4 already for the temperature of inlet exhaust gases delivered to the gas turbine T_2 higher than 1,400 K—Fig. 3.10.

However, the most important is to compare a compressor heat pump not to a power plant but to a combined heat and power plant [4], i.e. a source that co-generates heat and electricity. The conditions to make the use of a compressor heat pump thermodynamically effective are far more "sharp" than for condition (2.48). Their derivation (dependencies (2.57), (2.67)) are presented below.

For an independently operating compressor heat pump driven by the electric motor with a power of N_{HP} the consumption of fuel chemical energy used to its generation in the power plant with efficiency η_{el} is $\dot{E}_{ch}$—Fig. 2.7. The pump energy efficiency providing heating heat to consumers at amount $\dot{Q}^{HP}$ can be determined by using the equation:

$$\eta_{HP} = \frac{\dot{Q}^{HP}}{\dot{E}_{ch}} = \frac{\dot{Q}^{HP}}{N_{HP}} \frac{N_{HP}}{\dot{E}_{ch}} = \varepsilon_E \eta_{el} \tag{2.49}$$

By contrast, the efficiency of the combined heat and power plant featuring the same thermal efficiency $\dot{Q}^{HP}$ and electric power N_{CHP} (power N_{CHP} results from the used at the combined heat and power plant technology of cogeneration) and using the fuel chemical energy at amount $\dot{E}_{ch}^{CHP}$ described in the following equation:

$$\eta_{CHP} = \frac{\dot{Q}^{HP} + N_{CHP}}{\dot{E}_{ch}^{CHP}} \tag{2.50}$$

For the combined heat and power plant we should determine a partial efficiency for the very heat generation, because it provides its optimum energetic effectiveness. It is because heat is the main product of the cogeneration process, while electricity is its by-product. This is because of the fact that the demand for the product decides what the main product is, hence its location. The by-product, i.e. electricity, is generated as the main product in a different source. For the by-product in the cogeneration process such efficiency η_{el} of its generation should be treated as the one in the boundary process, i.e. in a power plant (generally speaking in a motor generating only electricity) that features the highest efficiency in a given country η_{el}. The use of fuel chemical energy needed to generate electricity (in other words, the energy that burdens its production) in the combined heat and power plant in a cogeneration process can be determined from formula:

$$\dot{E}_{ch}^{N_{CHP}} = \frac{N_{CHP}}{\eta_{el}} \tag{2.51}$$

Here, it should be emphasised that stream $\dot{E}_{ch}^{N_{CHP}}$ is the fuel "avoided" during the production of heat in the combined heat and power plant, while the revenue from selling electricity is, in turn, in economic calculation the avoided cost of its generation (3.28) [1, 3–10].

The use of fuel chemical energy burdening the heat production in the combined heat and power plant can be calculated from equation:

$$\dot{E}_{ch}^{\dot{Q}^{CHP}} = \dot{E}_{ch}^{CHP} - \dot{E}_{ch}^{N_{CHP}} \tag{2.52}$$

and the partial heat generation efficiency for the plant can be expressed by the following dependency:

$$\eta_p^{CHP} = \frac{\dot{Q}^{HP}}{\dot{E}_{ch}^{\dot{Q}^{CHP}}} = \frac{\dot{Q}^{HP}}{\dot{E}_{ch}^{CHP} - \frac{N_{CHP}}{\eta_{el}}} = \frac{\eta_{CHP}}{1 - \sigma\left(\frac{\eta_{CHP}}{\eta_{el}} - 1\right)} \tag{2.53}$$

wherein the index σ of cogeneration for the combined heat and power plant can be expressed by the following formula:

$$\sigma = \frac{N_{CHP}}{\dot{Q}^{HP}}. \tag{2.54}$$

Savings on natural fuel chemical energy used in the heat pump is determined by subtracting its consumption in the power plant used to generate power N_{HP} from the consumption in the combined heat and power plant replaced by the pump:

$$\Delta\dot{E}_{ch} = \dot{E} \to_{ch}^{\dot{Q}^{CHP}} - \dot{E}_{ch} = Q^{HP}\left(\frac{1}{\eta_p^{CHP}} - \frac{1}{\eta_{HP}}\right) \tag{2.55}$$

The savings on chemical energy would occur when the expression in brackets for formula (2.55) is positive. Therefore, it is sufficient to:

$$\Delta\dot{E}_{ch} = \dot{E}_{ch}^{\dot{Q}^{CHP}} \tag{2.56}$$

In order to make an independently operating pump thermodynamically viable as compared to the combined heat and power plant, the index ε_E has to follow the dependency:

$$\varepsilon_E > \frac{\eta_p^{CHP}}{\eta_{el}} = \frac{\eta_{CHP}}{\eta_{el}} \frac{1}{1-\sigma\left(\frac{\eta_{CHP}}{\eta_{el}}-1\right)} \tag{2.57}$$

The resulting from formula (2.57) value ε_E would have to be higher than its real value 2.5. Such a high value results from the fact that it depends not only on efficiency η_{el} (2.48), but also, and mainly, on the index value σ for cogeneration in the combined heat and power plant (2.57). The higher the index σ, the higher the partial efficiency for heat generation in the combined heat and power plant and thus the lower stream of fuel burnt in the plant that burdens its generation, so consequently the index ε_E of the heat pump has to be higher as well. Then, value ε_E must be high enough to make the stream of fuel burnt in the power plant (2.51) to produce electricity driving the pump not higher than the fuel stream related to heat generation in the combined heat and power plant (2.52). In practice, such a high index value ε_E that would guarantee the thermodynamic viability of the heat pump as compared to the combined heat and power plant is unavailable. The pump is even less viable in economic terms [4].

The thermodynamic ineffectiveness of using a compressor heat pump can be reduced by improving its cooperation with an electricity source—Fig. 3.2—i.e. by using not only electric power to drive it, but also by using the outlet heat of its exhaust gases—Fig. 2.8. Is it then possible that in this situation the pump could be thermodynamically viable? The answer is unequivocal—no (Sect. 2.5.1).

To sum up, it should be stated that the heating heat delivered by the pump is thermodynamically unviable. Furthermore, at the same time, it is quite expensive and would require subsidising. It results from a high annual cost of electricity needed to drive the pump and a high investment outlay on the pump. The cheapest heating heat can be directly obtained from engines generating heat in cogeneration with electricity [4]. At the same time its cost is many times lower than the cost of heat from the pump.

2.5.1 Thermodynamic Analysis of a Compressor Heat Pump System and A Hierarchical Gas–Gas Engine For Combined Heat and Electricity Production

As already mentioned, the thermodynamic ineffectiveness of using the compressor heat pump can be reduced by linking it with the hierarchical gas–gas engine not only by using electricity to drive it—Fig. 2.6—but also by using the heat exchanger *exhaust gas-central heating water and domestic hot water* and using exhaust gases from the engine—Fig. 2.8. At the same time, obviously, the heat of exhaust gases is not a heat source for the pump, as it is the ambience. Below we present a thermodynamic analysis of such a solution.

The use of engine exhaust gas heat, which should be clearly emphasised, does not increase the energy efficiency index ε_E of the very heat pump cycle, but it reduces the lower limit of the value for which it could turn out that it is viable for the

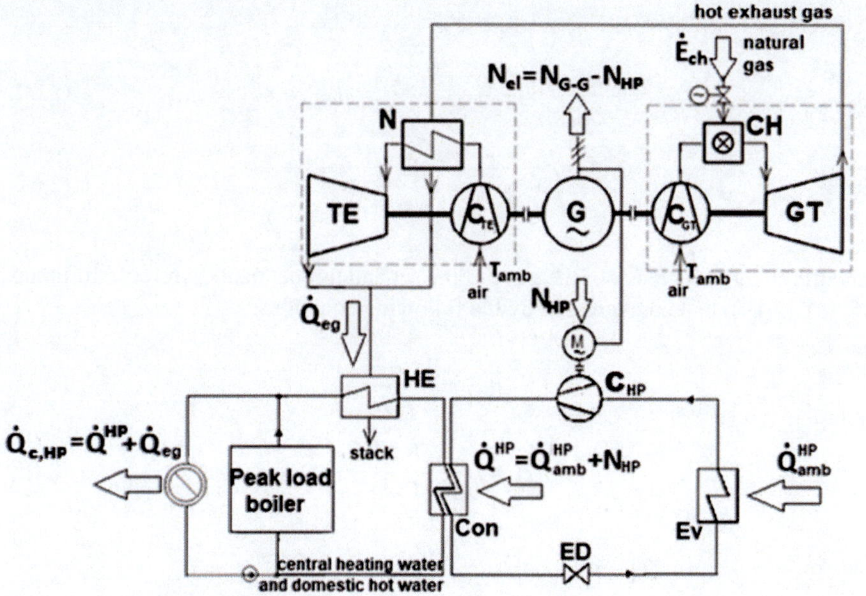

Fig. 2.8 The schematic diagram of the compressor heat pump and hierarchical gas–gas engine for cogeneration (combine production of heat and electricity): *G*—electric generator, *CH* —gas turbine combustion chamber, *M*—electric motor driving the compressor C_{HP} heat pump, *N*—air heater, C_{TE} —low-pressure compressor of turboexpander, C_{GT} —gas turbine high-pressure compressor, *GT* – gas turbine, *TE* – turboexpander, *HE* —heat exchanger exhaust gas-central heating water and domestic hot water, C_{HP} —heat pump compressor, *Ev*—heat pump evaporator, *Con*—heat pump condenser, *ED*—heat pump expansion device

thermodynamic reasons to replace the combined heat and power plant with a pump-fitted system presented in Fig. 2.8. However, it does not mean that also heat could turn out to be cheaper than the heat from the combined heat and power plant (Fig. 3.3, Sect. 3.3.1), as the cost of its production increases high investment outlays on the very pump and the lost revenue from selling electricity – the most precious energy form—that drives it. The lost revenue including capital costs and costs of maintenance and repairs resulting from outlays on the pump cannot compensate additional revenue from the increased heat production. Then, it should be emphasised that completely unbeatable in terms of price is the heat from the combined heat and power plant, generally speaking from engines generating heat in cogeneration with electricity [4].

The individual quantities presented in Fig. 2.8 can be expressed by the following formulas:

$$N_{G-G} = N_{GT} + N_{TE} \tag{2.58}$$

$$\dot{Q}_{eg} = (1 - \eta_{G-G})\eta_u \dot{E} \tag{2.59}$$

wherein:

$$\eta_{G-G} = \frac{N_{G-G}}{\dot{E}_{ch}} = \frac{N_{GT} + N_{TE}}{\dot{E}_{ch}} = \eta_{GT} + \eta_{TE} - \eta_{GT} \quad (2.60)$$

$$\varepsilon_E = \frac{\dot{Q}^{HP}}{N_{HP}} \quad (2.61)$$

and the energy efficiency of the gas turbine set and turboexpander (refer to formulas (2.18), (2.19)) can be expressed by the following equations:

$$\eta_{GT} = \frac{N_{GT}}{\dot{E}_{ch}} \quad (2.62)$$

$$\eta_{TE} = \frac{N_{TE}}{\dot{E}_{ch}(1 - \eta_{TG})} \quad (2.63)$$

where:

$\dot{E}_{ch}$	fuel chemical energy stream burnt in the gas turbine,
N_{HP}	driving power of the heat pump compressor,
N_{TE}	turboexpander power,
N_{GT}	power of the gas turbine set,
$\dot{Q}^{HP}, \dot{Q}_{eg}, \dot{Q}^{HP}_{amb}$	in sequence, heating heat streams emitted to the central heating water and domestic hot water system and by exhaust gases from the hierarchical gas–gas engine and the heat stream taken by the pump from ambience,
$\dot{Q}_{c,HP}$	heating heat stream generated in the compressor heat pump system and the hierarchical gas–gas engine,
η_u	level of usage of exhaust gas stream enthalpy from the hierarchical gas–gas engine.

Energy efficiency of the compressor heat pump system and hierarchical gas–gas engine can be expressed by the following formula:

$$\eta^{CHP}_{c,HP} = \frac{\dot{Q}_{c,HP} + N_{el}}{\dot{E}_{ch}} = \varepsilon_E\left(\eta_{G-G} - \frac{N_{el}}{\dot{E}_{ch}}\right) + (1 - \eta_{G-G})\eta_u + \frac{N_{el}}{\dot{E}_{ch}} \quad (2.64)$$

while the partial efficiency for heat generation in the system can be expressed by the following dependency:

$$\eta^{CHP}_{c,HP\,p} = \frac{\dot{Q}_{c,HP}}{\dot{E}^{\dot{Q}_{hw}}_{ch}} = \frac{\dot{Q}_{c,HP}}{\dot{E}_{ch} - \frac{N_{el}}{\eta_{el}}} = \frac{\varepsilon_E\left(\eta_{G-G} - \frac{N_{el}}{\dot{E}_{ch}}\right) + (1 - \eta_{G-G})\eta_u}{1 - \frac{N_{el}}{\eta_{el}\dot{E}_{ch}}} \quad (2.65)$$

where:

N_{el} electric power output from the compressor heat pump system and hierarchical gas–gas engine,

N_η energy efficiency of the so-called boundary power plant.

Using formula (2.56) and substituting there for η_{HP} partial efficiency $\eta_{c,\,HP\;p}^{CHP}$ of heat generation in the system presented in Fig. 2.8:

$$\eta_{c,HP\;p}^{CHP} > \eta_p^{CHP} \tag{2.66}$$

we receive a threshold dependency, i.e. minimum value of the energy efficiency index for the pump (refer to formula (2.57)), to make the system from Fig. 2.8 thermodynamically viable:

$$\varepsilon_E > \frac{\eta_p^{CHP}\left(1 - \frac{N_{el}}{\eta_{el}\dot{E}_{ch}}\right) - (1 - \eta_{G-G})\eta_u}{\eta_{G-G} - \frac{N_{el}}{\dot{E}_{ch}}} \tag{2.67}$$

Minimum index ε_E considering a small difference for quantities in the denominator of the formula (2.67) is very sensitive to value changes. It is also sensitive to changes in partial efficiency η_p^{CHP} (formula (2.53); in this formula there is a difference in the denominator) of generating heat in the combined heat and power plant. The efficiency, as already mentioned above, depends significantly on the power generation technology, i.e. on the index σ (2.54) of the combined heat and power plant cogeneration operation. For example, for the combined heat and power plant working in the *gas-steam* technology the index assumes the value significantly higher than for the combined heat and power plant where only the *Clausius-Rankine* cycle is used. Therefore, depending on the technology used the minimum values ε_E calculated from formula (2.67) that would guarantee the thermodynamic viability of using the pump in the system presented in Fig. 2.8 assume the same values. Furthermore, the values depend on the pump power N_{HP} that in these calculations is a pre-set value. For instance, when a *gas–gas* engine drives only the pump compressor, i.e. when $N_{HP} = N_{G-G}(N_{el} = 0)$, then:

$$\eta_{c,HP}^{CHP} = \eta_{c,HP\;p}^{CHP} = \varepsilon_E\eta_{G-G} + (1 - \eta_{G-G})\eta \tag{2.68}$$

and formula (2.67) reduces to the following form:

$$\varepsilon_E > \frac{\eta_p^{CHP} - (1 - \eta_{G-G})\eta_u}{\eta_{G-G}} \approx \frac{\eta_p^{CHP}}{\eta_{G-G}} \tag{2.69}$$

For example, if the system from Fig. 2.8 replaced the *gas-steam* combined heat and power plant, i.e. cogeneration plant, where combined *Joule-Brayton* and *Clausius-Rankine* cycles are used together in a hierarchical system, then the calculated from formula (2.69) minimum index value ε_E would have to be higher than approx. 7, while if it was supposed to replace the cogeneration plant where only *Clausius-Rankine* cycle is used, then the value ε_E would have to be higher than approx. 5.5. The values are unavailable for the heat pump.

Calculating the partial efficiency η_p^{CHP} of heat production in the *gas-steam* combined heat and power plant (2.53), the value of efficiency for the so-called boundary power plant while calculating electricity generation was obviously substituted by the efficiency of the *gas-steam* combined heat and power plant $\eta_{el} = 0.55$. It would be a mistake to substitute for those calculations the efficiency of the steam power plant $\eta_{el} = 0.45$ that would be obviously lower than the efficiency of the *gas-steam* power plant. The obtained then for $\eta_{el} = 0.45$ a higher value for the partial heat production would not result from the cogeneration, but from numerical manipulations.

Finally, it should be emphasised again that minimum values ε_E that would guarantee economic viability of using heat pumps are much higher than values ε_E calculated from formulas (2.57), (2.67) that would guarantee their thermodynamic viability (economic conditions for viability of pumps are presented in [4]). Therefore, heat pumps are not only thermodynamically unviable, but, first of all, they are economically unviable. Here, it is necessary to indicate that the economic effectiveness determines the suitability of using a specific technological solution, while the energetic (exergetic) analysis allows us to only look for the opportunities to improve the thermal processes used in thermodynamic systems.

References

1. Bartnik R (2009) Gas-steam power plants and combined heat and power plants. Energy and economic efficiency (in Polish: Elektrownie i elektrociepłownie gazowoparowe. Efektywność energetyczna i ekonomiczna), WNT, Warszawa (reprint 2012, 2017)
2. Kowalczyk T, Ziółkowski P, Badur J (2015) Exergy losses in the Szewalski binary vapor cycle. Entropy 17:7242–7265. https://doi.org/10.3390/e17107242
3. Bartnik R, Buryn Z, Hnydiuk-Stefan A (2017) Methodology and a continuous time mathematical model for selecting the optimum capacity of a heat accumulator integrated with a CHP plant (in Polish: Ekonomika Energetyki w modelach matematycznych z czasem ciągłym). Wydawnictwo Naukowe PWN, Warszawa
4. Bartnik R, Bartnik B (2014) Economic calculation in the energy sector (in Polish: Rachunek ekonomiczny w energetyce). WNT, Warszawa
5. Bartnik R, Buryn Z, Hnydiuk-Stefan A, Skomudek W (2020) Power plant retrofit and modernization (in Polish: Modernizacja elektrowni. Efektywność energetyczna i ekonomiczna), WNT, Warszawa
6. Bartnik R, Buryn Z, Hnydiuk-Stefan A, Skomudek W (2019) Dual-Fuel Gas-Steam Combined Heat and Power Plants (in Polish: Dwupaliwowe elektrownie i elektrociepłownie gazowoparowe. Efektywność energetyczna i ekonomiczna), WNT, Warszawa
7. Bartnik R, Buryn Z, Hnydiuk-Stefan A (2017) Investment strategy in heating and CHP. Mathematical Models, Springer, London
8. Bartnik R, Bartnik B, Hnydiuk-Stefan A (2016) Optimum investment strategy in the power industry. Mathematical Models, Springer, New York

9. Bartnik R (2013) The modernization potential of gas turbines in the coal-fired power industry. Thermal and Economic Effectiveness, Springer, London
10. Bartnik R, Buryn Z (2011) Conversion of coal-fired power plants to cogeneration and combined-cycle: thermal and economic effectiveness. Springer, London

Chapter 3
Thermodynamic and Economic Analysis of a Gas Turbine Set Coupled with a Turboexpander in a Hierarchical Gas–Gas System

3.1 Introduction

Analysing costs of electricity and heat production in hierarchical power plants and *gas-steam* [1] cogeneration plants (also referred to as *Combined Cycle Power Plants*) turns out that the capital component is a very significant factor of those costs. Investment outlays on a steam part using the *Clausius-Rankine* cycle consist approx. 40% of the outlays on *gas-steam* system, when the gas turbine (both gas and steam turbine have here more general meaning and cover proper turbines and all the necessary auxiliary devices) requires only 30% of those outlays. In addition, installation-construction work making up the remaining 30% of outlays, are mainly (over 2/3) outlays on the steam part-related installation. Consequently, unit (per unit of installed electric power) "turnkey" investment outlays on the so-called simple systems, i.e. power plants and combined heat and power plants using only the *Joule-Brayton* cycle (the so-called *Simple Cycle Power Plants*) are more than twice lower as compared to the outlays on combined systems and make up approx, 45% of those outlays [2]. Therefore, we should look for a way to reduce costs of heat and electricity production related to the steam part of hierarchical *gas-steam* power plants and combined heat and power plants. For example, in spite of installing the steam turbine in those systems one can consider using the turboexpander including the compressor and air heater—Figs. 3.1, 3.2 and 3.3. Then, the power plants and combined heat and power plants will use two *Joule-Brayton* cycles, the *Joule-Brayton* cycle of the gas turbine and the *Joule-Brayton* cycle of the turboexpander.

R. Bartnik and T. W. Kowalczyk, *Hierarchical Gas-Gas Systems*,
Power Systems, https://doi.org/10.1007/978-3-030-69205-6_3

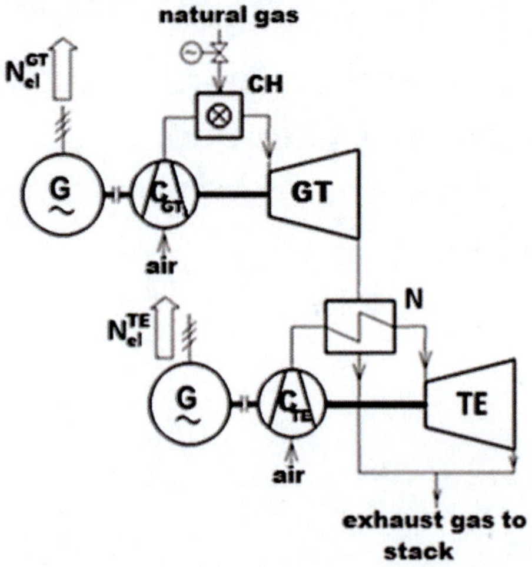

Fig. 3.1 The schematic diagram of the hierarchical system with the gas turbine and turboexpander in the two-shaft set-up used to generate electricity (G – electric generator, CH—gas turbine combustion chamber, N—air heater; the heater is a device that combines the gas turbine cycle with the turboexpander cycle, C_{TE} —low-pressure compressor, C_{GT} —high-pressure compressor, GT—gas turbine, TE—turboexpander)

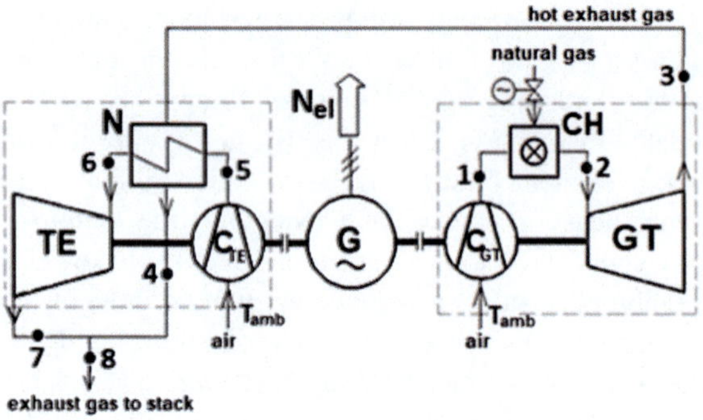

Fig. 3.2 The schematic diagram of the system with the gas turbine and turboexpander in the single-shaft set-up used to produce electricity; gas turbine set (GT + C_{GT}) with the turboexpander set (TE + C_{TE}) are installed on a common shaft and drive one, common generator G)

3.2 Thermodynamic Analysis of a Gas Turbine Generator Coupled with a Turboexpander

The *gas–gas* system can be also constructed in a two-shaft set-up—Fig. 3.1 [1].

However, in terms of investment the single-shaft system is cheaper—Figs. 3.2, 3.3. This system is patent-protected [3].

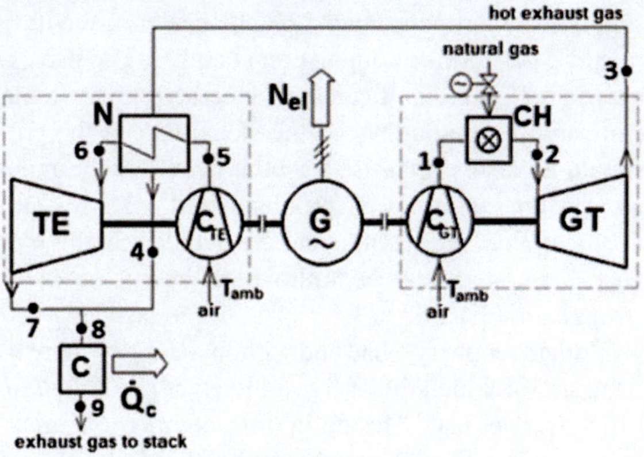

Fig. 3.3 The schematic diagram of the system with the gas turbine and turboexpander in the single-shaft set-up used to cogeneration (combined production of heat and electricity) (C —heat exchanger)

Dotted line rectangles in Fig. 3.2. and 3.3. mean that the turboexpander and the low-pressure compressor are integrated in a single casing and share a common shaft as is the case for the turbine part and gas turbine high-pressure compressor. This solution reduces investment outlays on the *gas–gas* engine.

The *gas–gas* hierarchical system is more expensive as compared to the gas turbine set operating with heat regenerative exchanger Fig. 3.4. However, such turbine sets with a capacity exceeding four megawatts are not produced [2]. Such low power results from construction problems caused by high differences in temperature between some turbine set components due to fitting inside them recovery heat exchangers. Meanwhile, systems with the turboexpander are not power restricted.

In thermodynamic calculations to compare the energy efficiency of the gas turbine set operating with heat recovery with the *gas–gas* system, it was assumed that the temperatures of exhaust gases delivered to the gas turbines T_2 in both cases were the same—Figs. 3.2, 3.4.

The turboexpander and compressor should be integrated in a single casing and share a common shaft, the same as it is in the case of gas turbines, where their turbine part is integrated with the compressor in one casing and installed on a single shaft, which significantly reduces the cost and price of this solution. The enthalpy of exhaust gases from the gas turbine would be used to heat air in the heater *N* that

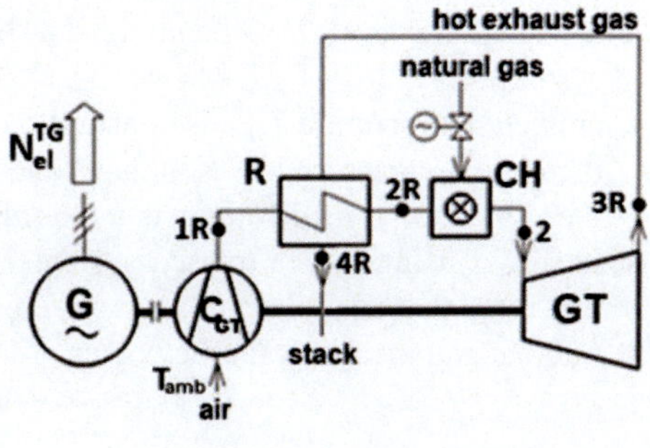

Fig. 3.4 The schematic diagram of the gas turbine set operating with heat regeneration (CH—gas turbine combustion chamber, R—regenerative heat exchanger, C_{GT} —high-pressure compressor, GT—gas turbine)

supplies the turboexpander *TE*. The heater in the turboexpander set would play the role of a gas turbine combustion chamber. The use of the turboexpander, as already mentioned, will significantly reduce investment outlays on the *gas–gas* system as compared to the outlays on the steam part of the *gas-steam* system. In addition, it should be also emphasised, which is extremely important, that in the case of *gas–gas* systems the relevant problems and costs related to water management of the *Clausius-Rankine* steam cycle are avoided, being a problem in *gas-steam* systems. *Gas–gas* systems can be built where there is no water, which should be once again strongly emphasised.

Both in recovery cycle and without recovery there is optimum ratio of heat transfer medium for which the energy efficiency η_E of *Joule-Brayton* cycle (formulas (3.2), (4.1)) is maximal. The optimum compression ratio value is z_{opt} the function of temperatures T_{amb}, T_2 mechanical efficiencies η_m^C, η_m^{GT} and internal of η_i^C,η_i^{GT} the compressor and turbine:

$$z_{opt} = \left(\frac{p_{opt}^{CH}}{p_{amb}}\right)^{\frac{\kappa-1}{\kappa}} = f\left(T_{amb}, T_2, \eta_m^{GT}, \eta_i^C, \eta_i^{GT}\right) \tag{3.1}$$

where:

p^{CH}, p_{amb}	heat transfer medium pressure in *Joule-Brayton* cycle in sequence during heat absorption and emission (p^{CH}– pressure behind the air compressor, p_{amb} – ambient pressure; for calculations it was assumed $p_{amb} = 0.1$ MPa),
T_{amb}	ambient temperature (for calculations it was assumed $T_{amb} = 288$ K),
T_2	temperature of exhaust gases delivered to the gas turbine,
κ	exponent of isentrope of the heat transfer medium (for calculations it was assumed $\kappa = 1.4$).

For thermodynamic calculations it was assumed that mechanical efficiencies of the compressor and gas turbine are equal and amount to $\eta_m^C = \eta_m^{GT} = \eta_m = 0.97$, while their internal efficiencies equal $\eta_i^{GT} = 0.87$, $\eta_i^C = 0.85$.

In addition, assuming that:

- the temperature of the medium flowing to the gas turbine T_2 has a constant value (temperature level T_2 depends on the heat-resistance of the gas turbine blade system)
- ambient temperature T_{amb} is constant
- there are no pressure losses in ducts and the heat exchanger
- the subject of considerations is a closed system, the heat transfer medium is an ideal gas using cycles (closed systems) in the analyses of real thermodynamic processes that take place in open systems significantly facilitate the analyses
- internal and mechanical efficiencies of machines have constant values,

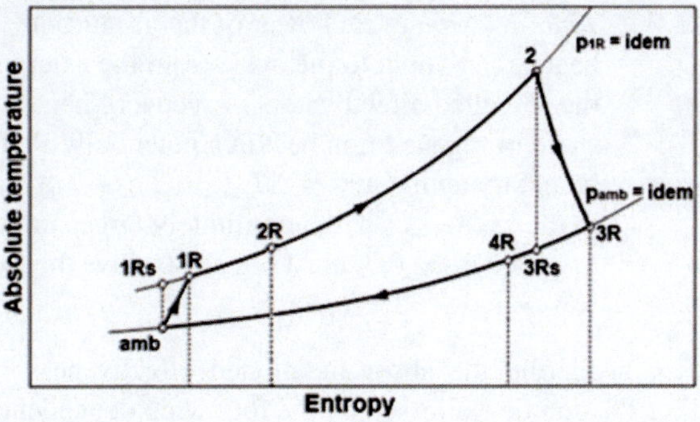

Fig. 3.5 The *Joule-Brayton* cycle of a gas turbine with heat regeneration (the heat of regeneration equals the surface area under the 3R-4R isobar section and the entropy axis; the surface area equals the surface area under the 1R-2R isobar section)

From the energy balance in the case of *Joule-Brayton* cycle with heat recovery—Figs. 3.4, 3.5—formula for its energy efficiency can be obtained [4]:

$$\eta_E^{GT} = \frac{N^{GT}}{\dot{Q}_{in}} = \frac{N_{i\ \exp} - N_{i\ C}}{\dot{Q}_{in}} = \frac{\eta_m(T_2 - T_{3R}) - \frac{1}{\eta_m}(T_{1R} - T_{ot})}{(T_2 - T_{2R})} \tag{3.2}$$

By substituting to (3.2) the temperature of the heat transfer medium (Fig. 3.5):

$$T_{1R} = T_{amb} + \frac{1}{\eta_i^S}(T_{1Rs} - T_{amb}) \tag{3.3}$$

$$T_{3R} = T_2 - \eta_i^{GT}(T_2 - T_{3Rs}) \tag{3.4}$$

$$T_{2R} = (1 - \eta_R)T_{1R} + \eta_R T_{3R} \tag{3.5}$$

$$T_{1Rs} = T_{amb} z^{GT} \tag{3.6}$$

wherein:

$$z^{GT} = \left(\frac{p_{1R}}{p_{amb}}\right)^{\frac{\kappa-1}{\kappa}} = \left(\frac{p_2}{p_{3R}}\right)^{\frac{\kappa-1}{\kappa}} \tag{3.8}$$

Equation (3.2) assumes the form

$$\eta_E^{GT} = \frac{N^{GT}}{\dot{Q}_{in}} = \frac{\eta_m \eta_i^{GT} T_2\left(1 - \frac{1}{z^{GT}}\right) - \frac{1}{\eta_m \eta_i^C} T_{amb}\left(z^{GT} - 1\right)}{T_2 - (1 - \eta_R)T_{amb}\left[1 + \frac{1}{\eta_i^C}\left(z^{GT} - 1\right)\right] - \eta_R T_2\left[1 - \eta_i^{GT}\left(1 - \frac{1}{z^{GT}}\right)\right]} \rightarrow \max \tag{3.9}$$

where:

N^{GT} *Joule-Brayton* cycle power of the gas turbine,

$\dot{Q}_{in}$ heat stream input to the cycle from the external source,

η_R the so-called effectiveness of recovery heat exchanger; defines as a percent share of regeneration heat in a potentially available heat within the range of boundary temperatures $\Delta T_{\max} = T_{3R} - T_{1R}$, the heat that could be obtained in the exchanger with an infinitely large surface area (for calculation it was assumed $\eta_R = 0.7$; must obviously have the following relations: $T_{3R} > T_{2R}$, $T_{4R} > T_{1R}$).

Considering the above-mentioned effectiveness η_R of the heat exchanger *R*, Fig. 3.4, can be expressed by the following dependency:

$$\eta_R = \frac{T_{3R} - T_{4R}}{T_{3R} - T_{1R}} = \frac{T_{2R} - T_{1R}}{T_{3R} - T_{1R}} \tag{3.10}$$

where:

T_{4R} temperature of exhaust gases from the recovery exchanger to the stack,

T_{2R} temperature of air behind the recovery device,

T_{1R} temperature of air behind the compressor,

T_{3R} temperature of exhaust gases from the gas turbine,

By substituting in formula (3.9) for η_R the value of zero, $\eta_R = 0$, we receive the formula for energy efficiency of *Joule-Brayton* cycle without recovery (Sect. 4.2.1, (4.1)).

In formula (3.10) it was assumed that exhaust gas thermal capacity and air streams delivered to the recovery device *R* (Fig. 3.4) are the same. This is because the analysis covers the cycle (Fig. 3.5), i.e. a closed system. In definition, the heat transfer medium thermal capacity stream is the same in its each point. This assumption applies also to a real system, because the exhaust gas mass from the gas turbine is almost the same as the stream delivered to the air regenerator. This is because the mass share of natural gas in exhaust gas created in the turbine combustion chamber CH is just approx. 2–3%, so the share of air makes up 97–98%.

After differentiating the formula (3.9) and using condition $d\eta_E^{GT}/dz^{GT} = 0$ for specific values, we obtain an T_{amb}, T_2, η_m, η_R, η_i^C, η_i^{GT} optimum compression ratio z_{opt}^{GT} (see Sect. 4.2.1, Eq. 4.3). Then, by substituting the value to (3.9) we can determine the maximum energetic efficiency of the gas turbine cycle $\eta_{E\max}^{GT}$. Knowing the value z_{opt}^{GT} from (3.1) we can also determine the value $p_{opt}^{CH,TG}$, by using irreversible adiabatic process we can calculate the temperature of the heat transfer medium behind the compressor T_1 and turbine T_3:

$$T_1 = T_{amb}\left[1 + \frac{1}{\eta_i^C}\left(z_{opt}^{GT} - 1\right)\right] \tag{3.11}$$

$$T_3 = T_2\left[1 - \eta_i^{GT}\left(1 - \frac{1}{z_{opt}^{GT}}\right)\right] \tag{3.12}$$

Using formula (3.10) we can determine the temperature of exhaust gases T_{4R} and air temperature T_{2R} behind the recovery heat exchanger—Fig. 3.4.

Here, it should be strongly emphasised that increasing the temperature of inlet air T_{amb} to the compressor is unfavourable, as it reduces the efficiency η_E^{GT}. For example, for the assumed values of machine and recovery exchanger efficiency, temperatures $T_{amb} = 288$ K and for example temperature $T_2 = 1{,}400$ K, the efficiency of cycle with recovery is $\eta_{E\max}^{R,GT} = 39.5\%$ at $z_{opt}^{R,GT} = 1.653$, for the cycle without recovery $\eta_{E\max}^{GT} = 31.9\%$ at $z_{opt}^{GT} = 2.214$. For temperature $T_{amb} = 298$ K $\eta_{E\max}^{R,GT} = 39\%$ at $z_{opt}^{R,GT} = 1.630$, $\eta_{E\max}^{GT} = 31\%$ and $z_{opt}^{GT} = 2.161$, while for temperature $T_{amb} = 278$ K $\eta_{E\max}^{R,GT} = 40.5\%$ and $z_{opt}^{R,GT} = 1.677$, $\eta_{E\max}^{GT} = 32.9\%$ and $z_{opt}^{GT} = 2.269$—Figs. 3.7 and 3.10.

Increasing η_R reduces $z_{opt}^{R,GT}$ and within the limit for $\eta_R = 1$ value $z_{opt}^{R,GT}$ assumes the value one, $z_{opt}^{R,GT} = 1$, which does not make a physical sense. As a result of overlapping isobars $p^{CH} = p_{amb}$ (3.1), the surface area of cycle representing in the case of irreversible adiabatic processes its power approximately N^{GT} assumes the value of zero, while at the same time the efficiency η_E^{GT} of *Joule-Brayton* cycle (3.9) achieves its maximum value.

By using formula (3.9) we can calculate also the optimum value z_{opt}^{TE} for the turboexpander cycle (compare to formula (4.43)). Then, in (3.9) for T_2 substitute the temperature of outlet exhaust gases in *GT* decreased by value ΔT_1: $T_6 = T_3 - \Delta T_1$. T_6 is the temperature of inlet air for the turboexpander—Figs. 3.2 and 3.3. The calculations assumed that the difference in temperature in the heat exchanger *N* between the temperature of T_3 the exhaust gas and the temperature T_6 of the air delivered to *TE* is $\Delta T_1 = 10\ K$. Adopting a lower value ΔT_1, even $\Delta T_1 = 0$, does not result in a large value increase z_{opt}^{TE}. It increases by just a few per mille, the same applies to efficiency $\eta_{E\max}^{TE}$.

By knowing optimum compression ratios for the cycle of the gas turbine z_{opt}^{GT} and turboexpander z_{opt}^{TE} we can, as a result, calculate the energy efficiency for the hierarchical *gas–gas* system (Sect. 2.2, formula (2.23)):

$$\eta^{G-G} = \eta_{E\max}^{GT} + \eta_{E\max}^{TE} - \eta_{E\max}^{GT}\eta_{E\max}^{TE} \tag{3.13}$$

Significant is the ratio of the turboexpander power N^{TE} to the gas turbine cycle N^{GT}. Keeping the same assumptions as used to determine dependency (3.9) the ratio can be determined from formula:

$$\frac{N^{TE}}{N^{GT}} = \frac{\dot{C}_{air}}{\dot{C}_{eg}} \frac{\eta_m\eta_i^{TE}T_6\left(1 - \frac{1}{z_{opt}^{TE}}\right) - \frac{1}{\eta_m\eta_i^{C,TE}}T_{amb}\left(z_{opt}^{TE} - 1\right)}{\eta_m\eta_i^{GT}T_2\left(1 - \frac{1}{z_{opt}^{GT}}\right) - \frac{1}{\eta_m\eta_i^{C,GT}}T_{amb}\left(z_{opt}^{GT} - 1\right)} \tag{3.14}$$

wherein the ratio of thermal capacities $\dot{C}_{air}/\dot{C}_{eg}$ of air and exhaust gas streams can be determined from energy balance for the air heater N—Figs. 3.2, 3.3. The heater is a device that combines the cycle of the gas turbine with the turboexpander cycle. The combination is performed by the exhaust *gas-air* system, where the low-temperature enthalpy of exhaust gases from the gas turbine is used to heat the air in the heater to deliver it to turboexpander:

$$\frac{\dot{C}_{air}}{\dot{C}_{eg}} = \frac{\dot{m}_{air}^{TE} c_{p,air}}{\dot{m}_{eg}^{TG} c_{p,eg}} = \frac{T_3 - T_4}{T_6 - T_5} = \frac{T_3 - T_5 - \Delta T_2}{T_3 - T_5 - \Delta T_1} \cong 1 \tag{3.15}$$

where:

$c_{p,air}$,$c_{p,eg}$ specific thermal capacities at a constant pressure of air and exhaust gas from the gas turbine, $c_{p,air} \cong c_{p,eg}$ (ratio of natural gas $\dot{m}_{gas}^{GT}$ and air streams $\dot{m}_{air}^{CH}$ delivered to the gas turbine combustion chamber is only approx. $\dot{m}_{gas}^{GT}/\dot{m}_{air}^{CH} \approx 2 \div 3\%$ and therefore it can be assumed that $c_{p,air} \cong c_{p,eg}$, because exhaust gases constitute almost "pure" air, obviously the mass balance is met $\dot{m}_{eg}^{GT} = \dot{m}_{gas}^{GT} + \dot{m}_{air}^{CH}$),

$\dot{m}_{air}^{TE}$, $\dot{m}_{eg}^{GT}$ mass streams of heat transfer media,

T_6 temperature of inlet air for the turboexpander, $T_6 = T_3 - \Delta T_1$,

T_4 temperature of exhaust gas from the air heater N, $T_4 = T_5 + \Delta T_2$, for calculations it was assumed $\Delta T_2 = 10\ K$,

T_5 temperature of outlet air from the compressor C_{TE} delivered to the heater N.

The value of ratio $\dot{C}_{air}/\dot{C}_{eg}$ (3.15) approximately equals one, because the difference $T_3 - T_5$ is significantly higher both from value ΔT_1 and ΔT_2. Increasing the difference in temperatures ΔT_1, ΔT_2 obviously reduces the surface area of the heat exchanger i.e. air heater N, so it reduces the financial outlays on it, and conversely, decreasing the difference raises the surface area and outlays. Furthermore, increasing the difference in temperature reduces the value of optimum compression ratio z_{opt}^{TE}, at the same time reducing the value of the turboexpander cycle power ratio N^{TE} to the power of gas turbine cycle N^{GT}, therefore it reduces the total power of *gas–gas* system and conversely. For example, for exhaust gas inlet temperature to the gas turbine of 1,400 K, increasing the difference in temperature from 10 to 30 K reduces the value z_{opt}^{TE} and N^{TE}/N^{GT}, respectively by 1.85% and 4.69%. Reducing the turboexpander power (gas turbine power obviously does not change), so reducing the total system power is relevant. Importantly, increasing the exhaust gas temperature delivered to the gas turbine decreases the reductions. And for the temperature of 1,800 K, the reductions of value z_{opt}^{TE} and N^{TE}/N^{GT} are only 1.67% and 2.77%. It is very relevant for this that the economic calculations (Sect. 3.3) assume unit investment outlays on the gas part related to the turboexpander equal to unit outlays on the gas part related to the gas turbine set, so the total investment outlays on the *gas–gas* system has been updated taking into account a relevant allowance. Thus, the economic calculations and the conclusions drawn from them showing the advantage of economic viability of *gas–gas* system over the *gas-steam* system are very "safe".

Formula (3.14) assumes that mechanical and internal efficiencies of compressors in both cycles as well as for the gas turbine and turboexpander are equal. Formula (3.15) assumes that the heat losses from heater to ambience amount to zero.

The temperatures of exhaust gases T_8 from *gas–gas* system to generate electricity—Fig. 3.2—is relatively high, over 200 °C—Fig. 3.7. So, it is high enough that wherever it is needed, the low-temperature enthalpy of those exhaust gases can be used to produce municipal heat in the heating exchanger—Fig. 3.3. This is because the temperatures needed for this heat are much lower than 200 °C as they range from 60 °C to 110 °C.

The total demand for municipal heating power from the combined heat and power plant used to heat, ventilate or air-condition rooms and to prepare domestic hot water changes throughout the year according to the standard summary diagram of demand for heating heat [1, 5]. During the heating season it depends on ambient temperature, in the post-heating one it is fixed and is needed only to heat domestic hot water.

In the case of industrial combined heat and power plants supplying heat for process purposes the temperatures needed for this heat are usually higher than 200 °C.

In the case of a *gas–gas* system combined operation—Fig. 3.3—a relevant quantity is the so-called cogeneration index for this operation. This is the ratio of electric power to thermal power. It can be calculated from formula:

$$\sigma = \frac{N_{el}^{GT} + N_{el}^{GT}}{\dot{Q}_C} = \frac{(N^{GT} + N^{TE})\eta_G}{\dot{Q}_C} = \frac{\left\{\eta_m(T_2 - T_3) - \frac{T_1 - T_{amb}}{\eta_m} + \frac{\dot{C}_{air}}{\dot{C}_{eg}}\left[\eta_m(T_6 - T_7) - \frac{T_5 - T_{amb}}{\eta_m}\right]\right\}\eta_G}{\left(1 + \frac{\dot{C}_{air}}{\dot{C}_{eg}}\right)(T_8 - T_9)\eta_{WC}} \tag{3.16}$$

where:

$\dot{Q}_C$ thermal power of the *gas–gas* system,
η_{WC} efficiency of the heating exchanger *C*—Fig. 3.3,
η_G efficiency of the electric generator *G*—Figs. 3.1, 3.2, 3.3 and 3.4.

Changes of index σ depend only on the changes in power value $\dot{Q}_C$. The electric power $N_{el}^{GT} + N_{el}^{TE}$ of the *gas–gas* system is constant, independent of $\dot{Q}_C$ (here it should be emphasised that a completely different problem are changes in the power of the gas turbine N^{GT} and turboexpander N^{TE} caused by the changes in air density resulting from the changes in ambient temperature). It is completely different for the *gas-steam* system, where the system electric power changes along with power changes $\dot{Q}_C$. This is because the steam turbine set changes in a very wide range as a result of high changes in relief heating streams from the set supplying heat exchangers. The range of changes in the steam turbine set exceeds 50% its maximum power [1]—Fig. 3.6.

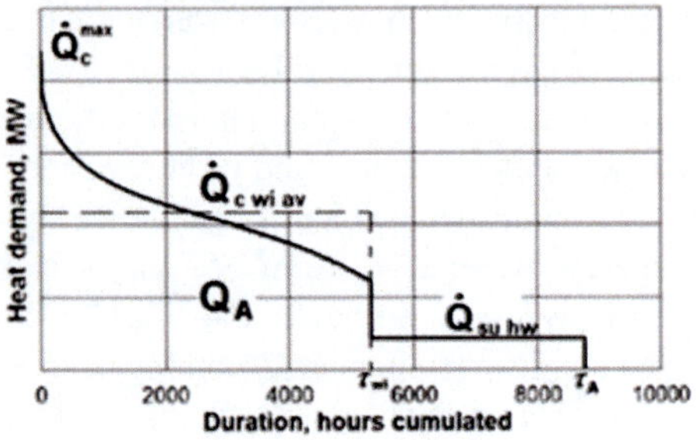

Fig. 3.6 Annual pre-arranged summary diagram of demand for municipal heating heat

A significant impact on the instantaneous value σ of the index has the temperature of exhaust gases from the heating exchanger C to the stack T_9—Fig. 3.3. This temperature changes depending on the thermal power $\dot{Q}_C$ of the *gas-gas* combined heat and power plant. The lowest value the index σ assumes for a minimum value of this temperature resulting from the minimum ambient temperature, i.e. when the combined heat and power plant has to operate at a maximum thermal power $\dot{Q}_C^{\max}$. By contrast, the temperature of inlet exhaust gas delivered T_8 to the exchanger is a constant value.

In economic calculations (Sect. 3.3) a relevant quantity is not the instantaneous value of the cogeneration index σ, but the value of the annual index, i.e. the ratio of annual electricity production $E_{el,A}$ to the annual heat production Q_A:

$$\sigma_A = \frac{E_{el,A}}{Q_A} \tag{3.17}$$

The value σ_A depends, obviously, on the combined heat and power cogeneration technology. For example, for municipal combined heat and power plants using the hierarchical *gas-steam* technology, the annual cogeneration index can achieve the value of even up to $\sigma_A^{G-P} \cong 4,1$ [1]. It should be emphasised that the value of the cogeneration index for instantaneous loads defined as a ratio of instantaneous electric power to the instantaneous thermal power $\sigma = N_{el}/\dot{Q}_c$(3.16) in municipal combined heat and power plants, i.e. working according to the pre-arranged diagram of demand for heating heat [1], Fig. 3.6, changes in time depending on ambient temperature; the value σ assumes the highest level in summer, i.e. when combined heat and power plants operate at a fixed minimum thermal power $\dot{Q}_c^{\min} = \dot{Q}_{suhw}$ only to heat the domestic hot water, and it assumes the lowest level σ when combined heat and power plants operate at a maximum thermal power $\dot{Q}_c^{\max}$.

The annual heat production Q_A in municipal combined heat and power plants is a definite integral of the curve of total demand for heating heat according to the annual pre-arranged diagram (so this corresponds to the surface area under the curve)—Fig. 3.6.

Heat Q_A can be expressed as a sum of average integral of thermal demands in the heating season (winter) and values of needs for domestic hot water in the post-heating season (summer) [1]:

$$Q_A = \dot{Q}_c^{\max}\tau_s = \dot{Q}_{cwi\ av}\Big|_0^{\tau_{wi}}\tau_{wi} + \dot{Q}_{suhw}\tau_{su} \tag{3.18}$$

therefore

$$\tau_s = \frac{\dot{Q}_{cwi\ av}\Big|_0^{\tau_{wi}}}{\dot{Q}_c^{\max}}\tau_{wi} + \frac{\dot{Q}_{suhw}}{\dot{Q}_c^{\max}}\tau_{su} = \xi\tau_{wi} + \beta\tau_{su} \tag{3.19}$$

where:

$\dot{Q}_c^{\max}$, $\dot{Q}_{cwi\ av}\Big\vert_0^{\tau_{wi}}$, $\dot{Q}_{suhw}$	in sequence, maximum (peak) thermal power, i.e. rated power of the combined heat and power plant, average power of the combined heat and power plant in the heating season and the power to prepare domestic hot water in the summer season,
τ_{su},τ_{wi}	duration times for respectively the post-heating season (summer) and heating season for the combined heat and power plant (for calculations it was assumed τ_{wi} = 5,200 h/a; $\tau_{su} = \tau_A - \tau_{wi}$, wherein the annual operation time τ_A of the combined heat and power plant was assumed to be τ_A = 8,424 h/a),
τ_s	annual usage time of the maximum power $\dot{Q}_c^{\max}$ of the combined heat and power plant (i.e. rated power, called peak power; for values assumed in the monograph β, ξ, τ_{wi}, τ_A time τ_s is 2,922 h/a).

Time τ_s for the municipal combined heat and power plants is only theoretical in nature and it shows only how many hours a year it would have to operate with its maximum thermal power $\dot{Q}_c^{\max}$ (i.e. peak power) to produce heating heat at an amount of Q_A. Introducing time τ_s and index σ_A is important, as they eliminate in the "economic" formula (3.27) quantities Q_A and $E_{el,A}$, whose ranges are very high, almost infinite. It significantly reduces the number of calculations of value $k_{h,av}^{G-G}$, $k_{el,av}^{G-G}$ to make it possible to show them in a graphical form.

In practice, for municipal combined heat and power plants value ξ is approx. 0.5, while $\beta \in \langle 0.05; 0.15\rangle$. The duration of the heating season τ_{wi} depending on a climatic zone (in Poland there are five zones) most frequently ranges from 5,040 $\div$ 5,400 h/a, i.e. 210 $\div$ 225 days. The summer time τ_{su} can be determined from the formula $\tau_A = \tau_{wi} + \tau_{su}$, where the annual operating time of the combined heat and power plant is on average $\tau_A \cong 8,424$ h/a (about 2 weeks of the combined heat and power plant downtime in summer holiday months). The annual time τ_s assumes the values from about 2,800 h/a to 3,000 h/a.

In the case of industrial combined heat and power plants their operation contrary to municipal combined heat and power plants is individual in its nature and features practically a constant thermal power. The annual times τ_s assume for them the values equal to times τ_A.

Formula (3.28) shows that the unit heat production cost depends mainly on the value of the annual cogeneration index σ_A, system energy efficiency η_c, unit investment outlays i and the sales price of generated electricity.

The annual value of cogeneration in the hierarchical *gas–gas* system in the case of operation in the combined heat and power plant to meet municipal heating demand according to the annual pre-arranged summary diagram of demand for heating heat [1, 5] can be determined from formula:

$$\sigma_A = \frac{E_{el,A}}{Q_A} = \frac{\left(N_{el}^{GT} + N_{el}^{TE}\right)\tau_A}{\dot{Q}_c^{\max}\tau_s} = \frac{\tau_A}{\tau_s}\sigma_{\min} \tag{3.20}$$

wherein:

$$N_{el}^{GT} + N_{el}^{TE} = \sigma_{\min}\dot{Q}_c^{\max} \tag{3.21}$$

Value $\sigma_{\min}$ can be determined from formula (3.16) for a minimum temperature of outlet exhaust gases T_9 from the heating exchanger C—Fig. 3.3. Then, heating heat stream $\dot{Q}_c$ assumes the maximum value $\dot{Q}_c \equiv \dot{Q}_c^{\max}$, which at a constant electric power $N_{el}^{GT} + N_{el}^{TE}$ causes that the cogeneration index assumes the minimum value $\sigma = \sigma_{\min}$. For calculations it was assumed that the minimum temperature value was $T_9 = T_9^{\min} = 353$ K.

Another significant quantity is the energy effectiveness for heat and electricity generation in the hierarchical *gas–gas* system. It can be determined from formula:

$$\eta_c = \frac{N_{el}^{GT} + N_{el}^{TE} + \dot{Q}_c}{\dot{Q}_{fuel}} = \frac{\left\{\eta_m(T_2 - T_3) - \frac{T_1 - T_{amb}}{\eta_m} + \frac{\dot{C}_{air}}{\dot{C}_{eg}}\left[\eta_m(T_6 - T_7) - \frac{T_5 - T_{amb}}{\eta_m}\right]\right\}\eta_G}{T_2 - T_1} + \frac{\left(1 + \frac{\dot{C}_{air}}{\dot{C}_{eg}}\right)(T_8 - T_9)\eta_{WC}}{T_2 - T_1} \tag{3.22}$$

The efficiency η_c depends on the temperature T_9 of exhaust gases from the heating exchanger C to the stack. In the heating and post-heating season its value can be determined from equations, respectively:

$$\xi = \frac{\dot{Q}_{cwi\ av}\Big|_0^{\tau_{wi}}}{\dot{Q}_c^{\max}} = \frac{T_8 - T_9^{wi\ av}}{T_8 - T_9^{\min}} \tag{3.23}$$

$$\beta = \frac{\dot{Q}_{suhw}}{\dot{Q}_c^{\max}} = \frac{T_8 - T_9^{su}}{T_8 - T_9^{\min}} \tag{3.24}$$

wherein for calculations it was assumed that $\xi = 0.5$ and $\beta = 0.1$. Then by using temperatures $T_9 = T_9^{wi\ av}$ and $T_9 = T_9^{su}$ from formula (3.22) are calculated value η_c^{wi} and η_c^{su}, respectively.

To calculate the unit cost of heat production by using formula (3.27), we should substitute for the value of combined heat and power plant energy effectiveness η_c its annual value $\eta_c = \eta_{c,A}$. The value can be expressed by the following formula:

$$\eta_{c,A} = \frac{(N_{el}^{GT} + N_{el}^{TE})\tau_A + Q_A}{\dot{Q}_{fuel}\tau_A} = \\ = \frac{\left\{\eta_m(T_2 - T_3) - \frac{T_1 - T_{amb}}{\eta_m} + \frac{\dot{C}_{air}}{\dot{C}_{eg}}\left[\eta_m(T_6 - T_7) - \frac{T_5 - T_{amb}}{\eta_m}\right]\right\}\eta_G}{T_2 - T_1}\frac{1 + \sigma_A}{\sigma_A}. \tag{3.25}$$

The annual fuel consumption $\dot{Q}_{fuel}\tau_A$ in formula (3.27) can be expressed by formula (3.25), but also by using the dependency (refer to formula (3.9)):

$$\dot{Q}_{fuel}\tau_A = \frac{N^{GT}}{\eta_{E\max}^{GT}}\tau_A \tag{3.26}$$

Figures 3.7, 3.8, 3.9, 3.10, 3.11 and 3.12 present the results of multi-variant thermodynamic calculations performed for the gas turbine with and without heat recovery and for the hierarchical *gas–gas* system, i.e. for the gas turbine without

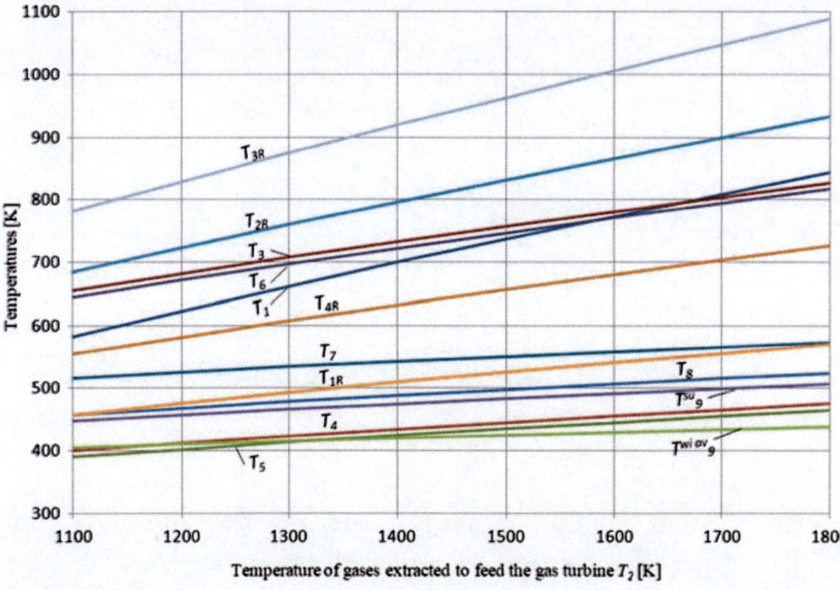

Fig. 3.7 Temperature values T_1,T_{1R},T_{2R},T_3,T_{3R},T_4,T_{4R},T_5,T_6,T_7,T_8,$T_9^{wi\ av}$,T_9^{su} as the function of the temperature of exhaust gases delivered to the gas turbine T_2

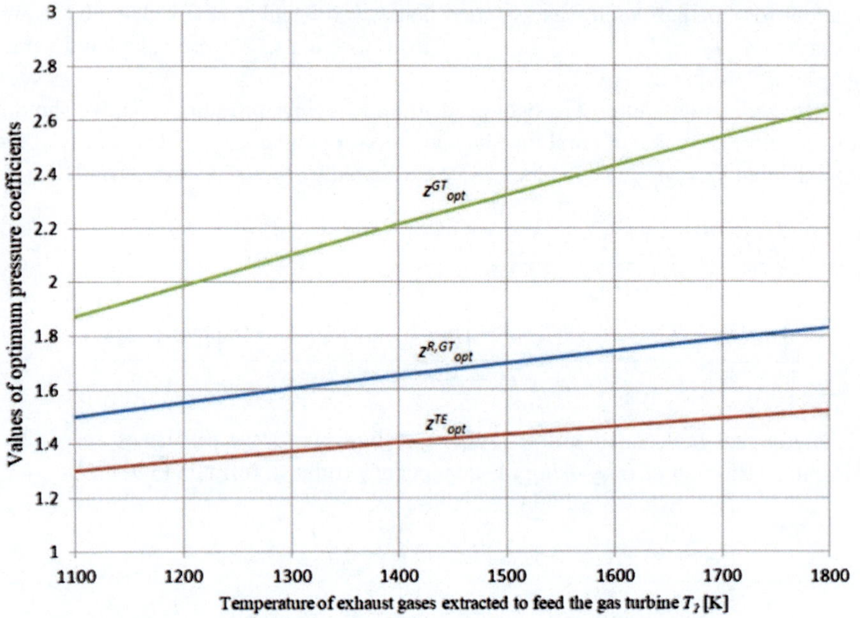

Fig. 3.8 Values of optimum compression ratios as z_{opt}^{GT}, z_{opt}^{TE}, $z_{opt}^{R,GT}$ the function of the temperature of exhaust gases delivered to the gas turbine T_2

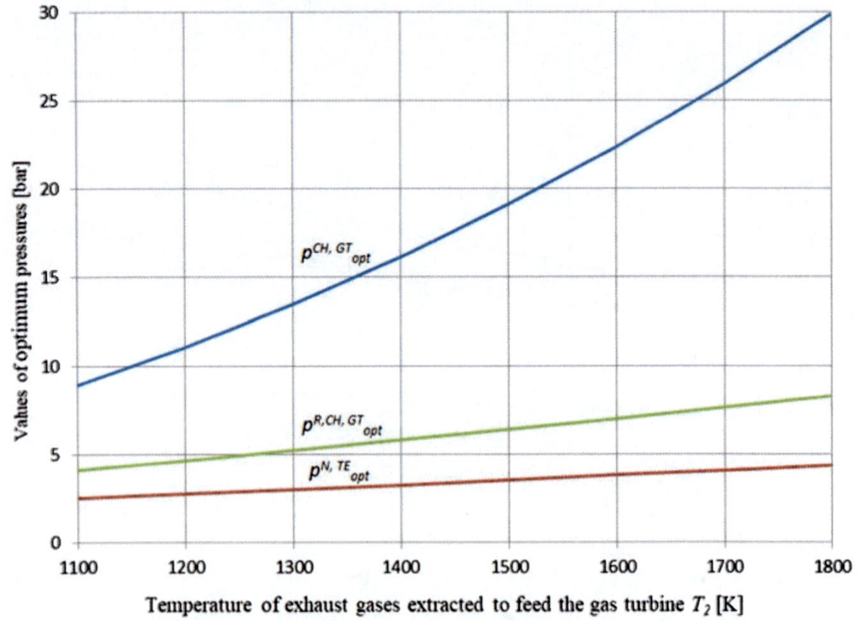

Fig. 3.9 Values of optimum pressures $p_{opt}^{CH,GT}$, $p_{opt}^{N,TE}$, $p_{opt}^{R,CH,GT}$ as the function of the temperature of exhaust gases delivered to the gas turbine T_2

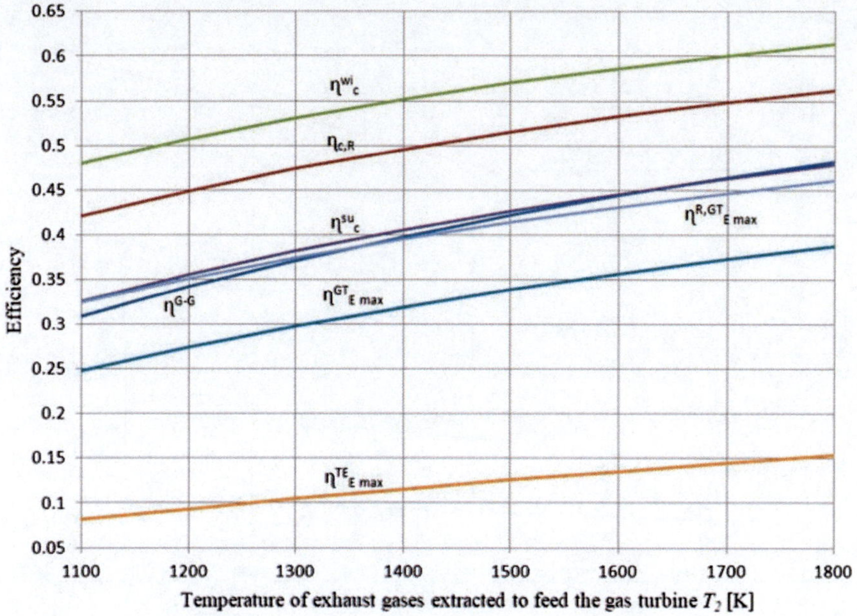

Fig. 3.10 Efficiency values η^{G-G}, $\eta_{c,R}$, η_c^{wi}, η_c^{su}, $\eta_{E\,\max}^{GT}$, $\eta_{E\,\max}^{TE}$, $\eta_{E\,\max}^{R,GT}$ as the function of the temperature of exhaust gases delivered to the gas turbine T_2

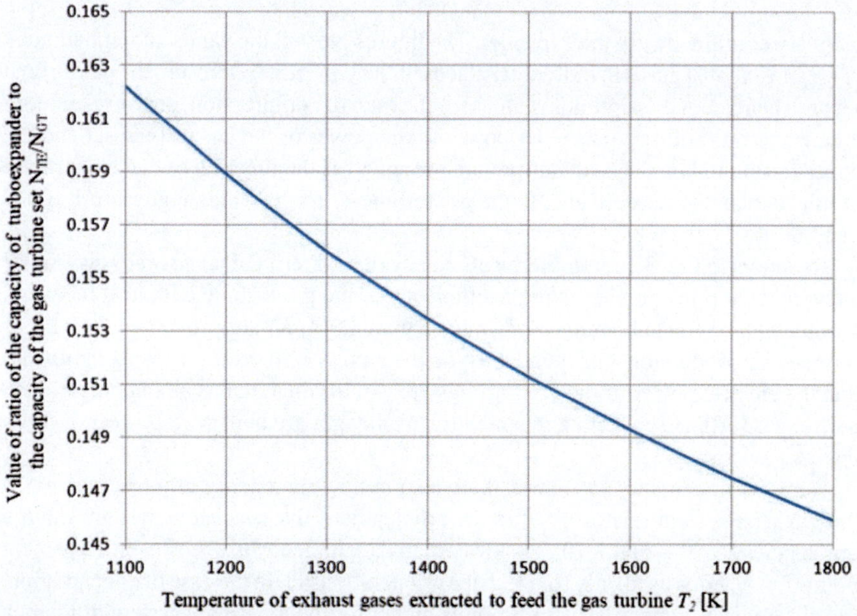

Fig. 3.11 Values of turboexpander power ratio to the power of the gas turbine set N^{TE}/N^{GT} as the function of the temperature of exhaust gases delivered to the gas turbine T_2

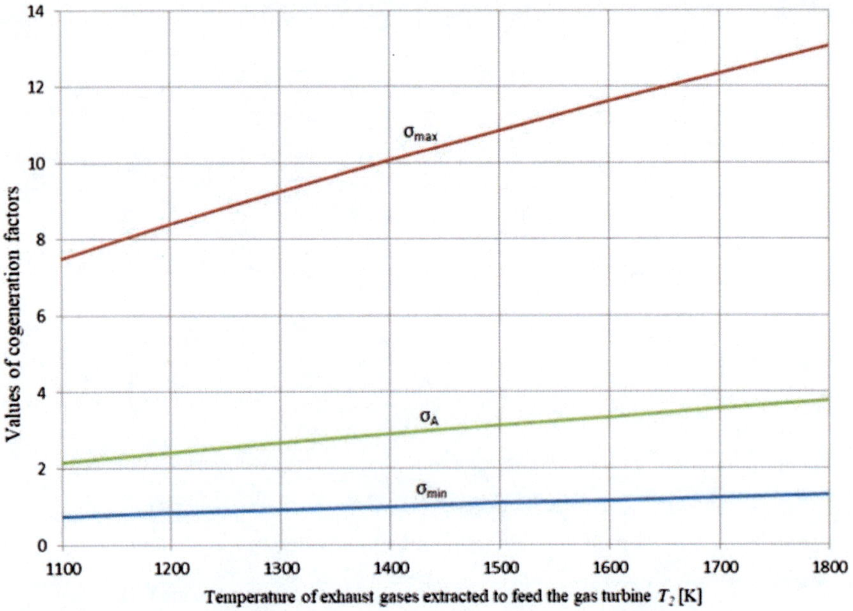

Fig. 3.12 Values of cogeneration rates $\sigma_{\min}$, $\sigma_{\max}$, σ_A as the function of the temperature of exhaust gases delivered to the gas turbine T_2

recovery combined with the expander. The figures present the values of temperatures of air and exhaust gases in individual places of the *gas–gas* system and the gas turbine set with heat recovery, optimum values for the ratio of compression, pressures, energy efficiency, ratio of turboexpander power to the power of the gas turbine set and the cogeneration indexes. The quantities are presented as the function of the temperature of inlet exhaust gases delivered to the gas turbine T_2 for its wide range from 1,100 K to 1,800 K.

As shown in Fig. 3.10, the energy efficiency of the hierarchical *gas–gas* cycle is by a few percent points higher than the efficiency of the gas turbine with heat recovery. Its takes place for temperatures T_2 higher than 1,300 K, so in practice for all gas turbines in production, independently of their power. In addition, it is significant that the electric power of the *gas–gas* system is unlimited, but as already mentioned above, gas turbine sets with a regeneration exchanger are built with powers of just a few megawatts.

Here, it is necessary to answer the question how the energy efficiency of η^{G-G} the *gas–gas* system relates η^{G-S} to the efficiency of the *gas-steam* system. Energy efficiencies η^{G-G}—Fig. 3.10.—is lower than the efficiency η^{G-S} by even 10 percent points for systems used only for electricity generation [1]. In the case of cogeneration, the difference is almost 30 percent points higher for the *gas-steam* system, for which the cogeneration efficiency ranges from about 80% to 85% [1], while for the *gas–gas* system it reaches the value of about 55%—Fig. 3.10. The higher efficiency of *gas-steam* systems is caused by the fact that they are fitted with the heat recovery steam generator (HRSG) and provide the capacity to use the low-temperature enthalpy of exhaust gases from the gas turbine all year round. The temperature of exhaust gases from the heat recovery steam generator, both for cogeneration and purely condensation of *gas-steam* systems, has a low value for the whole year, ranging from 70 °C to 90 °C, while in the *gas–gas* system the temperature is relatively high. For the system generating only electricity it is about $t_8 \in \langle 180;\ 250^{\circ}\mathrm{C}\rangle$, while for the cogeneration in the heating season it is about $t_9^{wiav} \in \langle 130;\ 160^{\circ}\mathrm{C}\rangle$ and in the post-heating season it is about $t_9^{su} \in \langle 170;\ 240^{\circ}\mathrm{C}\rangle$—Fig. 3.7. To reduce the temperature, it is possible to use the low-temperature enthalpy of the outlet exhaust gases from the *gas–gas* system, as driving heat for the thermal absorption chiller. The low-temperature enthalpy of exhaust gases from the gas turbine in the *gas-steam* system is used in the heat recovery steam generator to produce steam supplying the steam turbine and to heat exhaust condensate from the steam turbine condenser. It takes place in an integrated recovery boiler at its final zone in the exchanger of low-pressure regeneration. Enthalpy can be also used in the dearation evaporator built in the boiler [1, 6]. The use of exhaust gas enthalpy in the boiler for such a low temperature of exhaust gas increases the electricity production in the steam turbine set as compared to the production in the turboexpander in the *gas–gas* system. The ratio of expander power to gas turbine power N^{TE}/N^{GT}—Fig. 3.11.—is even more than 3 times lower than the ratio of N^{ST}/N^{GT} steam turbine power to gas turbine power in the *gas-steam* system in the case of the system only for electricity generation, while for cogeneration the value of the ratio is approx. 2 times lower.

3.3 Economic Analysis of a Gas Turbine Generator Coupled with a Turboexpander

The thermodynamic analysis allows us to find an opportunity to improve technology processes and construction solutions for machines and devices. In a market economy it is the economic criterion, profit and its maximisation that decide whether a specific technical solution should be used, it is the economic viability that makes the basis for making a decision. This is because the economic criterion is superior to the technical criterion. However, it should be emphasised that the economic analysis is possible only after a previous thermodynamic analysis. Its results make up input values for the economic analysis.

Important. The input values for the economic analysis of a gas–gas system include: gas turbine power N^{GT}, turboexpander power N^{TE}, thermal power $\dot{Q}_c$ of the heating exchanger and fuel chemical energy stream of gas $\dot{E}_{ch}$ used in the gas turbine. Here, it should be strongly emphasised that the power of the gas turbine determines all the remaining values: $N^{TE} = f(N^{GT})$, $\dot{Q}_c = f(N^{GT})$ and, which is obvious $\dot{E}_{ch} = f(N^{GT})$. Thus, all the listed power values of the gas–gas system depend only on the power of the gas turbine and the inlet temperature of exhaust gases T_2 (temperature level depends on the heat-resistance of the gas turbine blade system). Another important issue is to develop a mathematical model of economic viability of a gas–gas system operation to use them to generalise the considerations, i.e. to use the calculation results for any gas turbine power N^{GT}, i.e. any power of the gas–gas system. To achieve this the model was provided with quantities without specific values: N^{TE}/N^{GT}, $\dot{Q}_c/(N^{GT})$ and $\dot{E}_{ch}/N^{GT}$. Only this "dimensionless" approach allows us, which should be strongly emphasised again, to generalise the considerations, i.e. evaluate the economic viability of the gas–gas system with any power: both electrical and thermal. Therefore, the results presented in the chapter are universal, in other words, they refer to any system power. The power values, as already mentioned, and which should be emphasised again, depend on the power of the gas turbine and T_2 temperature of exhaust gases delivered to it from the CH combustion chamber of the gas turbine.

3.3.1 Unit Heat Production Costs in a Gas–Gas and Gas-Steam Systems

Unit heat production cost both for the *gas-gas* combined heat and power plant as well as the *gas-steam* combined heat and power plant can be determined from the discounted profit *NPV* achieved during *T* years of operation [7–10] (in the case of the *gas-steam* system as expander power N^{TE} in formulas (3.27), (3.28) substitute the average annual steam turbine power N^{ST}):

$$
\begin{aligned}
NPV = \Bigg\{ & \left(N^{GT} + N^{TE}\right)\eta_G(1-\varepsilon_{el})\tau_A \frac{e_{el}^{t=0}}{a_{el}-r}[e^{(a_{el}-r)T}-1] + Q_A \frac{e_c^{t=0}}{a_c-r}[e^{(a_c-r)T}-1] + \\
& - \dot{Q}_{fuel}\tau_A(1+x_{sw,m,was})\frac{e_{fuel}^{t=0}}{a_{fuel}-r}[e^{(a_{fuel}-r)T}-1] + \\
& - \dot{Q}_{fuel}\tau_A \frac{\rho_{CO_2} p_{CO_2}^{t=0}}{a_{CO_2}-r}[e^{(a_{CO_2}-r)T}-1] - \dot{Q}_{fuel}\tau_A \frac{\rho_{CO} p_{CO}^{t=0}}{a_{CO}-r}[e^{(a_{CO}-r)T}-1] + \\
& - \dot{Q}_{fuel}\tau_A \frac{\rho_{NO_X} p_{NO_X}^{t=0}}{a_{NO_X}-r}[e^{(a_{NO_X}-r)T}-1] - \dot{Q}_{fuel}\tau_A \frac{\rho_{SO_2} p_{SO_2}^{t=0}}{a_{SO_2}-r}[e^{(a_{SO_2}-r)T}-1] + \\
& - \dot{Q}_{fuel}\tau_A \frac{\rho_{dust} p_{dust}^{t=0}}{a_{dust}-r}[e^{(a_{dust}-r)T}-1] - \dot{Q}_{fuel}\tau_A(1-u)\frac{\rho_{CO_2} e_{CO_2}^{t=0}}{b_{CO_2}-r}[e^{(b_{CO_2}-r)T}-1] + \\
& -(1+x_{sal,t,ins})J(1-e^{-rT})\frac{\delta_{serv}}{r} - zJ[\frac{1-e^{-rT}}{T}+1]\Bigg\}(1-p)
\end{aligned}
\tag{3.27}
$$

therefore for $NPV = 0$ and $a_c = 0$ by using formulas (3.20) and (3.26) we get a dependency for average heat production cost in the period of T years:

$$
\begin{aligned}
\kappa_{h,av} = {} & \frac{\sigma_A r}{\left(1+\frac{N^{TE}}{N^{GT}}\right)\eta^{GT}_{E\max}\eta_G(1-e^{-rT})}\left\{\frac{(1+x_{sw,m,was})e^{t=0}_{fuel}}{a_{fuel}-r}[e^{(a_{fuel}-r)T}-1]+\right.\\
& +\frac{\rho_{CO_2}p^{t=0}_{CO_2}}{a_{CO_2}-r}[e^{(a_{CO_2}-r)T}-1]+\frac{\rho_{CO}p^{t=0}_{CO}}{a_{CO}-r}[e^{(a_{CO}-r)T}-1]+\\
& +\frac{\rho_{NO_X}p^{t=0}_{NO_X}}{a_{NO_X}-r}[e^{(a_{NO_X}-r)T}-1]+\frac{\rho_{SO_2}p^{t=0}_{SO_2}}{a_{SO_2}-r}[e^{(a_{SO_2}-r)T}-1]+\\
& \left.+\frac{\rho_{dust}p^{t=0}_{dust}}{a_{dust}-r}[e^{(dust-r)T}-1]+(1-u)\frac{\rho_{CO_2}e^{t=0}_{CO_2}}{b_{CO_2}-r}[e^{(b_{CO_2}-r)T}-1]\right\}+\\
& +(1+x_{sal,t,ins})\frac{i\sigma_A\delta_{serv}}{\tau_A}+\frac{\sigma_A rzi}{\tau_A(1-e^{-rT})}\left(\frac{1-e^{-rT}}{T}+1\right)\\
& -\frac{(1-\varepsilon_{el})r\sigma_A e^{t=0}_{el}}{(a_{el}-r)(1-e^{-rT})}[e^{(a_{el}-r)T}-1]
\end{aligned}
\tag{3.28}
$$

where:

$a_{el}, a_{fuel}, a_{CO_2}, a_{CO}, a_{SO_2}, a_{NO_X}, b_{CO_2}$ — exponents illustrating changes in time for electricity prices, fuel, environmental fees, purchase of CO_2 emission permissions,

$e^{t=0}_{el}, e^{t=0}_{fuel} e^{t=0}_{CO_2}, p^{t=0}_{CO_2}$ etc. — initial values of prices of electricity, fuel, purchase of CO_2 emission permissions and environmental fees,

e_{CO_2} — prices of carbon dioxide emission permissions, PLN/Mg_{CO_2},

i — unit (per power unit) investment outlay on the combined heat and power plant, $i = J/(N^{GT}_{el} + N^{TE}_{el})$, PLN/MW (value i depends on the technology of combined heat and electricity production and electric power installed),

J — investment outlays,

$p_{CO_2}, p_{CO}, p_{NO_x}, p_{SO_2}, p_{dust}$ — unit rate per CO_2, CO, NO_x, SO_2, dust emission, PLN/Mg,

p — income tax rate for gross profit,

Q_A — heat production per annum,

r — investment capital interest rate J,

T — combined heat and power plant operating period in years,

u — share of fuel chemical energy in its total annual consumption, for which the purchase of required CO_2 emission permissions purchase,

$x_{sw,m,was}$ — coefficient taking into account the cost of top-up water, auxiliary materials (chemicals), waste disposal.

$x_{sal,t,ins}$ — coefficient taking into account the cost of pays, taxes, insurance policies, etc.

z discounting coefficient (freezing ratio) of investment capital J at the moment of investment end,

ε_{el} nindex of own electrical demands of the power unit,

δ_{serv} nannual fixed cost rate depending on investment outlays (costs of maintenance, repairs of devices).

$\rho_{CO_2}, \rho_{CO}, \rho_{NO_x}, \rho_{SO_2}, \rho_{dust}$ CO_2, CO, NO_x, SO_2, dust emission per fuel chemical energy unit, kg/GJ,

For the *gas–gas* system as a coefficient $x_{sw,m,was}$ taking into account top-up water, chemicals for its treatment and waste water disposal, enter zero, $x_{sw,m,was} = 0$, for the *gas-steam* system the value was assumed to be $x_{sw,m,was} = 0.02$.

Figures 3.13, 3.14, 3.15, 3.16, 3.17, 3.18, 3.19, 3.20 and 3.21 present the results of multi-variant calculations of unit heat production cost in *gas–gas* and *gas-steam*. combined heat and power plants. As unit investment outlays in the calculations presented for the *gas–gas* combined heat and power plant the value $i = 350$ US/kW was substituted, while for the *gas-steam* one $i = 780$ US/kW was substituted [2]. The outlays do not cover combined heat and power plants or power plants (Figs. 3.22 and 3.23) with electric power ranging from about 80 MW to100 MW. Here, it should be emphasised that outlays i show a hyperbolic decline with power increase above 80–100 MW and, conversely, increase when power drops below 80–100 MW.

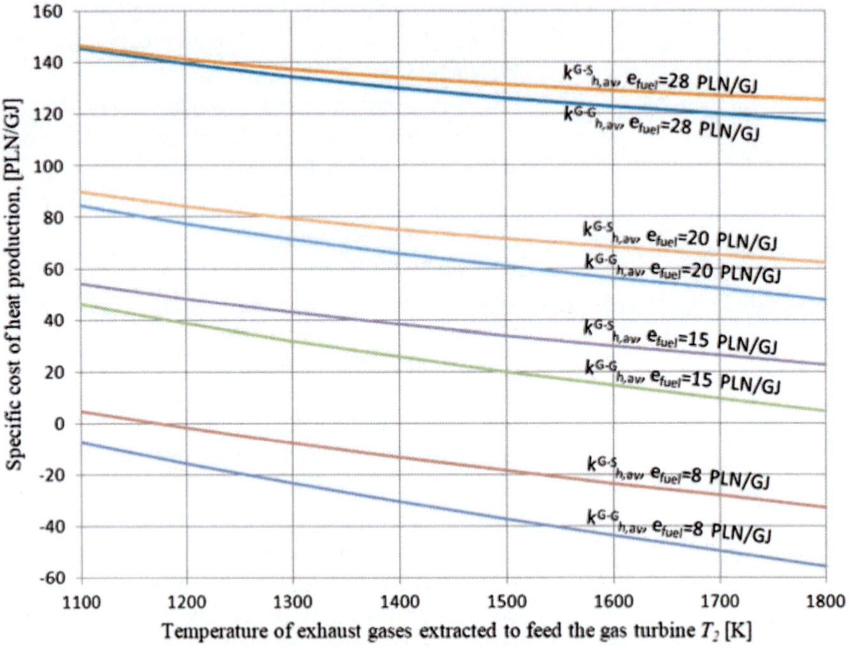

Fig. 3.13 The unit costs of heat production $k_{h,av}^{G-G}$, $k_{h,av}^{G-S}$ in gas–gas and gas-steam systems as the function of the temperature of exhaust gases delivered to the gas turbine T_2 for the electricity price amounting to 150 PLN/MWh (values $k_{h,av}^{G-S}$ calculated for the ratio value of $N^{ST}/N^{GT} = 0.28$)

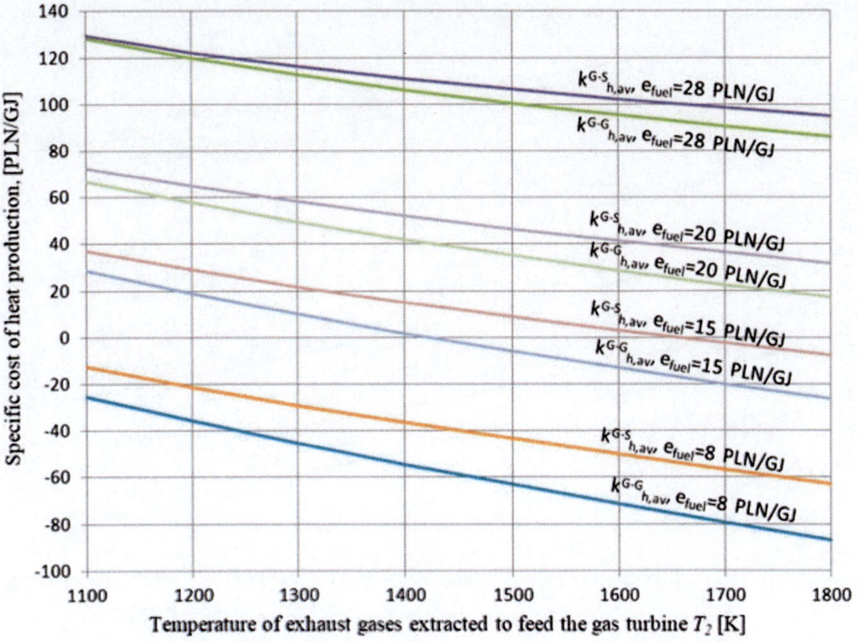

Fig. 3.14 The unit costs of heat production $k_{h,av}^{G-G}$, $k_{h,av}^{G-S}$ in gas–gas and gas-steam systems as the function of the temperature of exhaust gases delivered to the gas turbine T_2 for the electricity price amounting to 180 PLN/MWh (values $k_{h,av}^{G-S}$ calculated for the ratio value of $N^{ST}/N^{GT} = 0.28$)

The above-presented calculations show that the *gas–gas* system is more economically viable than the *gas-steam* system. The viability is the higher the higher is the temperature of inlet exhaust gases delivered to the gas turbine T_2, hence the higher is the energy efficiency of the gas turbine (Fig. 3.10).

3.3.2 Unit Electricity Production Costs in a Gas–Gas and Gas-Steam System

The unit electricity production cost both for the *gas-gas* power plant and *gas-steam* power plant can be determined from formula (3.27) substituting in it $Q_A = 0$ and then for $NPV = 0$ and $a_{el} = 0$ we get a formula for the average annual cost of electricity production in the period of T years (as already mentioned above, in the case of the *gas-steam* system for the power of the expander N^{TE} in formula (3.29) enter the power of steam turbine N^{ST}):

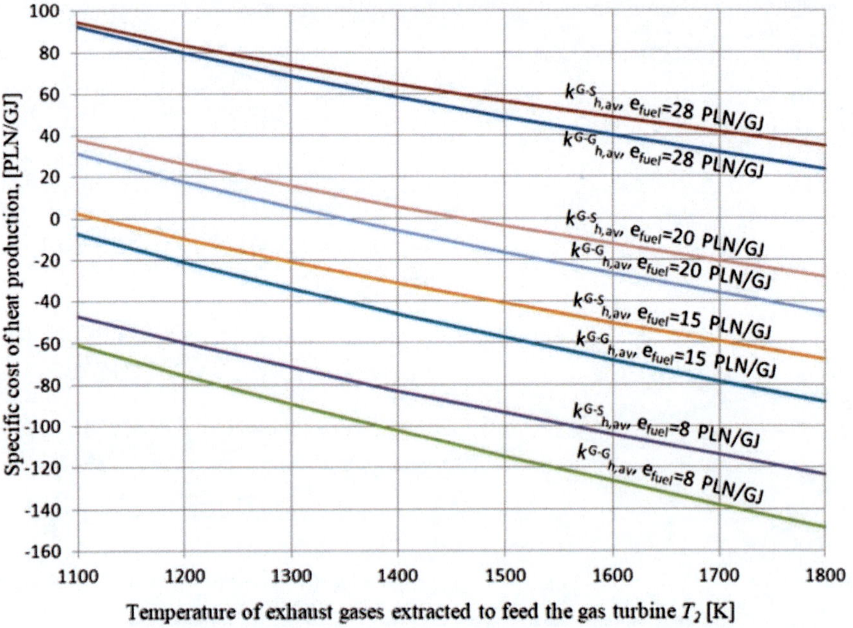

Fig. 3.15 The unit costs of heat production $k_{h,av}^{G-G}$, $k_{h,av}^{G-S}$ in gas–gas and gas-steam systems as the function of the temperature of exhaust gases delivered to the gas turbine T_2 for the electricity price amounting to 240 PLN/MWh (values $k_{h,av}^{G-S}$ calculated for the ratio value of $N^{ST}/N^{GT} = 0.28$)

$$
\begin{aligned}
k_{el,av} = & \frac{r}{\left(1+\frac{N^{TE}}{N^{GT}}\right)\eta_{E\max}^{GT}\eta_G(1-\varepsilon_{el})(1-e^{-rT})} \\
& \left\{(1+x_{sw,m,was})\frac{e_{fuel}^{t=0}}{a_{fuel}-r}[e^{(a_{fuel}-r)T}-1]+\right. \\
& +\frac{\rho_{CO_2}p_{CO_2}^{t=0}}{a_{CO_2}-r}[e^{(a_{CO_2}-r)T}-1]+\frac{\rho_{CO}p_{CO}^{t=0}}{a_{CO}-r}[e^{(a_{CO}-r)T}-1]+ \\
& +\frac{\rho_{NO_X}p_{NO_X}^{t=0}}{a_{NO_X}-r}[e^{(a_{NO_X}-r)T}-1]+\frac{\rho_{SO_2}p_{SO_2}^{t=0}}{a_{SO_2}-r}[e^{(a_{SO_2}-r)T}-1]+ \\
& \left.+\frac{\rho_{dust}p_{dust}^{t=0}}{a_{dust}-r}[e^{(a_{dust}-r)T}-1]+(1-u)\frac{\rho_{CO_2}e_{CO_2}^{t=0}}{b_{CO_2}-r}[e^{(b_{CO_2}-r)T}-1]\right\}+ \\
& +(1+x_{sal,t,ins})\frac{i\delta_{serv}}{(1-\varepsilon_{el})\tau_A}+\frac{riz}{(1-\varepsilon_{el})(1-e^{-rT})\tau_A}\left(\frac{1-e^{-rT}}{T}+1\right)
\end{aligned}
\tag{3.29}
$$

Figures 3.22 and 3.23 present the results of multi-variant calculations of unit electricity production costs $k_{el,av}^{G-G}$, $k_{el,av}^{G-S}$ in *gas–gas and gas-steam* systems as the function of the temperature of exhaust gases delivered to the turbine T_2. In Fig. 3.22 values $k_{el,av}^{G-S}$ calculated for the ratio $N^{ST}/N^{GT} = 0.5$, in Fig. 3.23. for $N^{ST}/N^{GT} = 0.55$. The above-presented calculations show that the *gas-steam* system is, though only to a small extent, more economically viable than the *gas–gas* system. This advantage decreases with a decrease in gas prices. However, it should be strongly

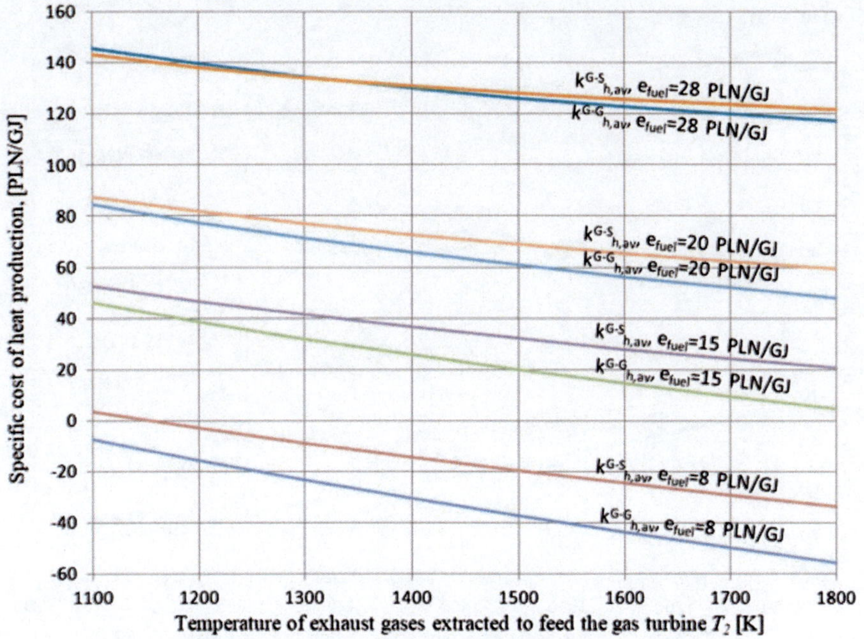

Fig. 3.16 The unit costs of heat production $k_{h,av}^{G-G}$, $k_{h,av}^{G-S}$ in gas–gas and gas-steam systems as the function of the temperature of exhaust gases delivered to the gas turbine T_2 for the electricity price amounting to 150 PLN/MWh (values $k_{h,av}^{G-S}$ calculated for the ratio value of $N^{ST}/N^{GT} = 0.3$)

emphasised again that the *gas–gas* system can be installed in places without water, which makes a strong advantage over the *gas-steam* system.

3.4 Summary and Final Conclusions

The calculation performed and presented in Figs. 3.13, 3.14, 3.15, 3.16, 3.17, 3.18, 3.19, 3.20 and 3.21 show that hierarchical *gas-gas* combined heat and power plants are more economically viable than hierarchical *gas-steam* combined heat and power plants. In both combined heat and power plants the power of gas turbine sets is the same. There are two reasons for higher viability of *gas–gas* combined heat and power plants. The first one consists in over two times lower unit investment outlays (per unit of electric power installed), i.e. over two times lower capital costs and operating costs determined by those outlays [8–10]. The second reason for this is that decreasing the annual electricity production in the *gas–gas* combined heat and power plant as compared to the *gas-steam* combined heat and power plant is so small that smaller revenue from its sales is compensated in excess by the above-mentioned lower annual

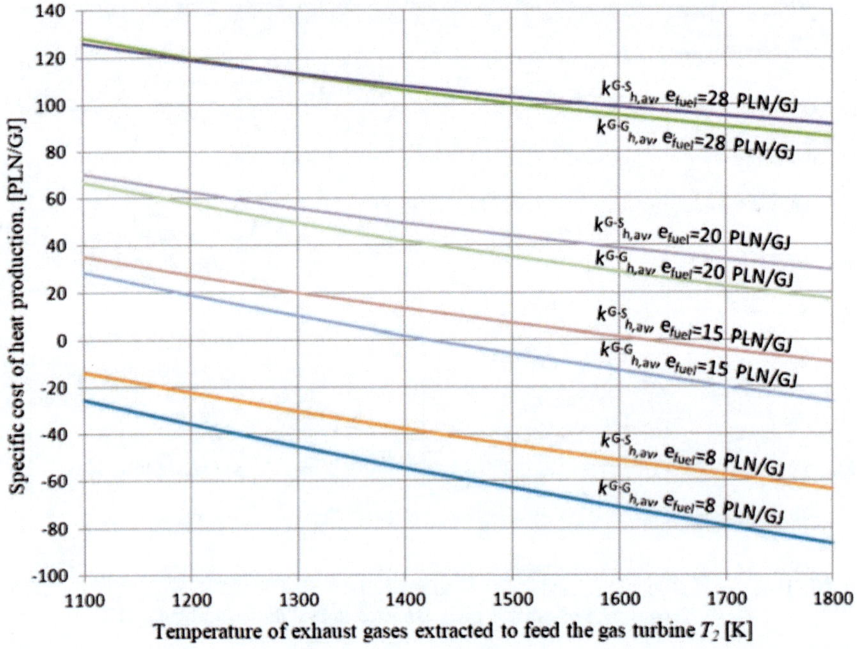

Fig. 3.17 The unit costs of heat production $k_{h,av}^{G-G}$, $k_{h,av}^{G-S}$ in gas–gas and gas-steam systems as the function of the temperature of exhaust gases delivered to the gas turbine T_2 for the electricity price amounting to 180 PLN/MWh (values $k_{h,av}^{G-S}$ calculated for the ratio value of $N^{ST}/N^{GT} = 0.3$)

capital and operating costs. At the same time, the higher the cost of electricity, the higher is the economic viability of *gas–gas* combined heat and power plants.

However, slightly higher, by a few percent points, is the economic viability of *gas-steam* power plants as compared to *gas–gas* power plants—Figs. 3.22 and 3.23. This results from higher electricity production in *gas-steam* power plants as compared to their operation in cogeneration systems. This is because heating steam is not collected from steam turbine relief valves to supply heating exchangers. As a result, the electricity production in the steam turbine set for its fully condensation-based operation is more than 3 times higher than the electricity production in the turboexpander in the *gas-gas* power plant, while for operation in cogeneration systems, it is only about 2 times higher. Obviously, this applies to the same power of gas turbines sets in both systems.

Considering the above-mentioned, the efficiency of electricity production in *gas-steam* systems is always higher than in *gas–gas* systems. Higher efficiency of those systems actually depends on the isotherm (which is also an isobar), when the heat of steam condensation is output to the atmosphere from *Clausius-Rankine* cycle. The isotherm almost corresponds to the ambient isotherm T_{amb} for the most thermodynamically ideal *Carnot* engine—Fig. 1.1. The heat transfer medium emitting heat to the atmosphere in the *gas–gas* system in the isobaric process of "low" *Joule-Brayton*

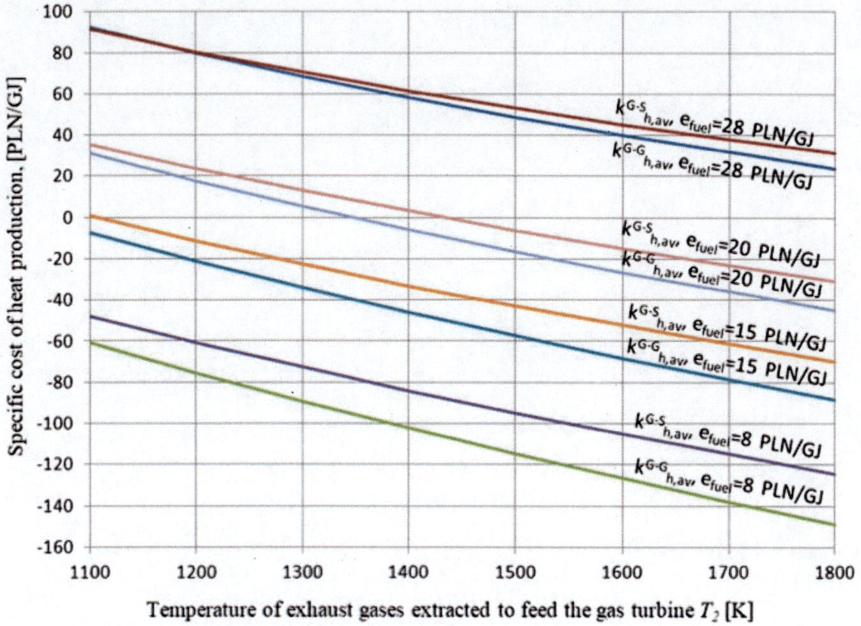

Fig. 3.18 The unit costs of heat production $k_{h,av}^{G-G}$, $k_{h,av}^{G-S}$ in gas–gas and gas-steam systems as the function of the temperature of exhaust gases delivered to the gas turbine T_2 for the electricity price amounting to 240 PLN/MWh (values $k_{h,av}^{G-S}$ calculated for the ratio value of $N^{ST}/N^{GT} = 0.3$)

cycle, i.e. the turboexpander cycle, has a significantly T_{amb} entropic average temperature (Chap. 2). As a result, as already mentioned above, the ratio of turboexpander power to the ratio of gas turbine power N^{TE}/N^{GT} in the *gas–gas* system is even more than 3 times lower compared to the ratio of steam turbine power to gas turbine power N^{ST}/N^{GT} in the *gas-steam system* at the same power in both gas turbine sets. This is the case when only electricity is generated in both systems. In the case of cogeneration operation the ratio of N^{TE}/N^{GT} is only about 2 times lower than N^{ST}/N^{GT}. The higher electricity production in the *gas-steam* system is available thanks to the integrated heat recovery steam generator (HRSG) that makes it possible to use the low-temperature enthalpy of exhaust gases from the gas turbine to generate steam supplying the steam turbine. The temperature of exhaust gases from the recovery boiler, both for cogeneration and purely condensation of *gas-steam* systems, has a low value for the whole year, ranging from 70 °C to 90 °C, while in *gas–gas* systems the temperature of exhaust gases is relatively high. For systems producing only electricity (the *C* heating exchanger does not occur in Fig. 3.3), the temperature of exhaust gases delivered to the stack ranges from about 180 °C to 250 °C (the temperature depends on the exhaust gas temperature supplying the gas turbine), while for cogeneration in the heating season ranges from about 130 °C to 160 °C and in the post-heating season ranges from about 170 °C to 240 °C. Obviously, they include the temperatures of exhaust gas downstream the *C* heating exchanger—Fig. 3.3.

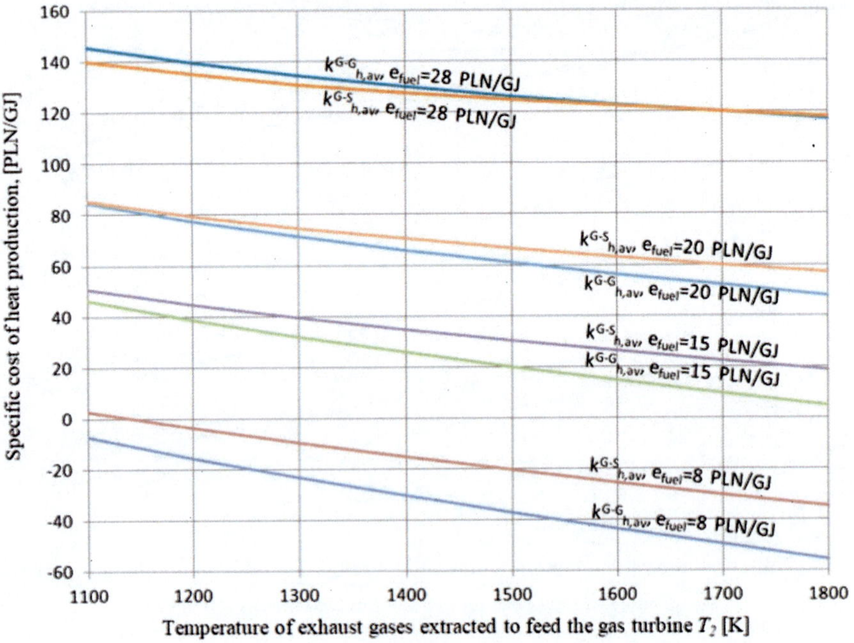

Fig. 3.19 The unit costs of heat production $k_{h,av}^{G-G}$, $k_{h,av}^{G-S}$ in gas–gas and gas-steam systems as the function of the temperature of exhaust gases delivered to the gas turbine T_2 for the electricity price amounting to 150 PLN/MWh (values $k_{h,av}^{G-S}$ calculated for the ratio value of $N^{ST}/N^{GT} = 0.32$)

By contrast, importantly, as already mentioned above, the economic viability of the *gas–gas* system operating in cogeneration is higher than the viability of the hierarchical *gas-steam* combined system, while economic *gas–gas* and *gas-steam* systems producing only electricity are comparable in spite of lower electricity production in the *gas–gas* system. This advantageous economic viability of *gas–gas* engines, which should be emphasised again, results from significantly lower unit investment outlays (per unit of electric power installed). They constitute about 45% of unit outlays on *gas-steam* systems.

There is another important matter. The efficiency of *gas–gas* η^{G-G} engines (3.13) for contemporary gas turbines, i.e. for high temperatures T_2 (Fig. 3.10), is comparable to the efficiency of steam engines for fresh steam supercritical parameters, where only *Clausius-Rankine* cycle is used. At the same time, unit financial outlays on *gas–gas* engines are a few times lower. It can be expected that in spite of expensive natural gas, more than twice more expensive per fuel chemical energy unit than coal, they will be more economically viable.

To make the summary complete, it is necessary, which is extremely important, to *expressis verbis* say that *gas–gas* systems can be built in places with no water, which is their great advantage that should not be overestimated.

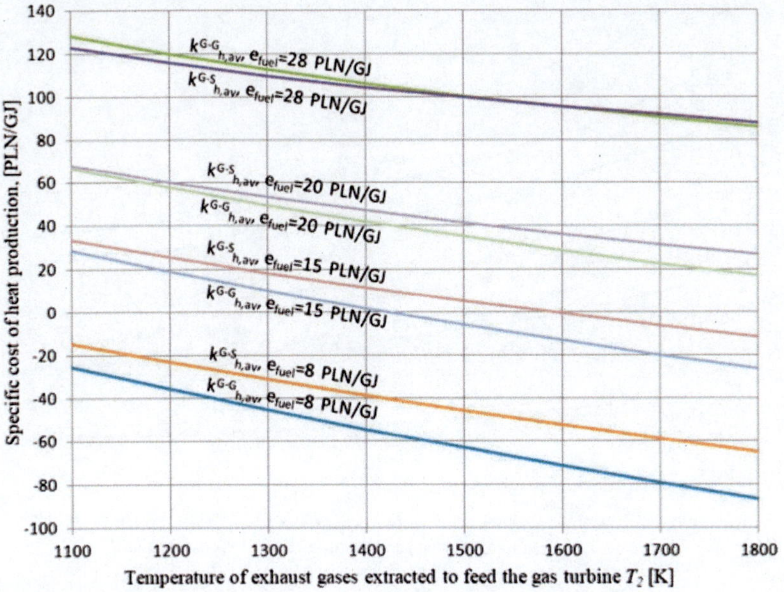

Fig. 3.20 The unit costs of heat production $k_{h,av}^{G-G}$, $k_{h,av}^{G-S}$ in gas–gas and gas-steam systems as the function of the temperature of exhaust gases delivered to the gas turbine T_2 for the electricity price amounting to 180 PLN/MWh (values $k_{h,av}^{G-S}$ calculated for the ratio value of $N^{ST}/N^{GT} = 0.32$)

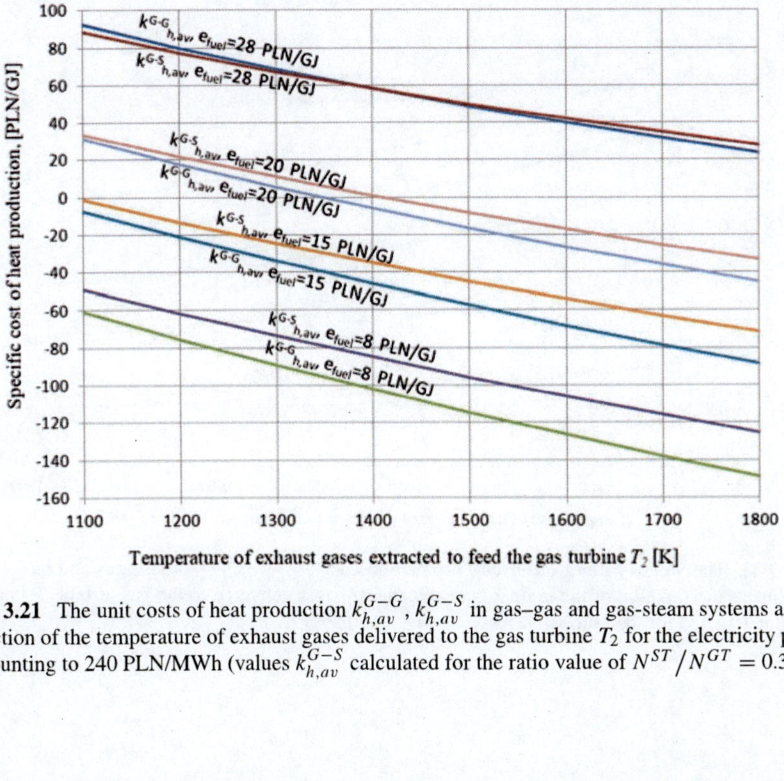

Fig. 3.21 The unit costs of heat production $k_{h,av}^{G-G}$, $k_{h,av}^{G-S}$ in gas–gas and gas-steam systems as the function of the temperature of exhaust gases delivered to the gas turbine T_2 for the electricity price amounting to 240 PLN/MWh (values $k_{h,av}^{G-S}$ calculated for the ratio value of $N^{ST}/N^{GT} = 0.32$)

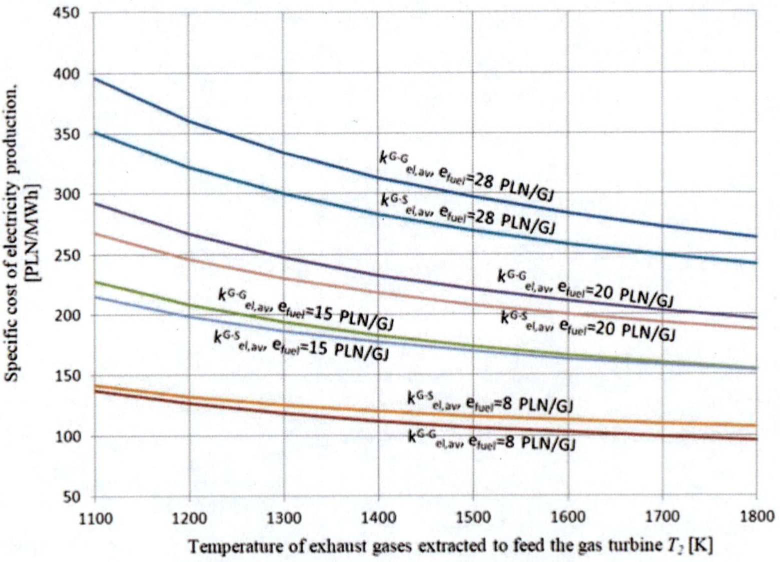

Fig. 3.22 The values of unit electricity production costs $k^{G-G}_{el,av}$, $k^{G-S}_{el,av}$ in gas–gas and gas-steam systems as the function of the temperature of exhaust gases delivered to the gas turbine T_2(values $k^{G-S}_{el,av}$ calculated for the ratio value of $N^{ST}/N^{GT} = 0.5$)

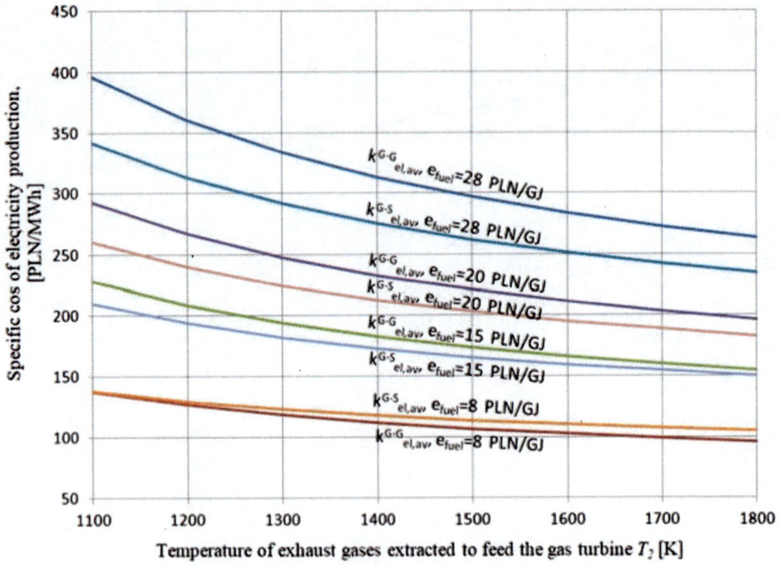

Fig. 3.23 The values of unit electricity production costs $k^{G-G}_{el,av}$, $k^{G-S}_{el,av}$ in gas–gas and gas-steam systems as the function of the temperature of exhaust gases delivered to the gas turbine T_2(values $k^{G-S}_{el,av}$ calculated for the ratio value of $N^{ST}/N^{GT} = 0.55$)

In addition, the calculation results presented in the Figs. 3.13, 3.14, 3.15, 3.16, 3.17, 3.18, 3.19, 3.20, 3.21, 3.22 and 3.23 confirm the general truth that the cheaper the fuel, the more economically viable are cheap power generation installations.

References

1. Bartnik R (2009) Gas-steam power plants and combined heat and power plants. Energy and economic efficiency (in Polish: Elektrownie i elektrociepłownie gazowoparowe. Efektywność energetyczna i ekonomiczna), WNT, Warszawa (reprint 2012, 2017).
2. Gas Turbine World (2007–2008) GTW Handbook. Volume 26, Pequot Publication, Inc. Southport, USA
3. Bartnik R (2013) System for generating electricity and heat. Patent P.389820, 2013
4. Szargut J (1998) Technical thermodynamics (in Polish: Termodynamika Techniczna). Wydawnictwo Politechniki Śląskiej, Gliwice
5. Bartnik R, Buryn Z (2011) Conversion of coal-fired power plants to cogeneration and combined-cycle: thermal and economic effectiveness. Springer, London
6. Bartnik R, Buryn Z, Hnydiuk-Stefan A, Skomudek W (2019) Dual-Fuel Gas-Steam Combined Heat and Power Plants (in Polish: Dwupaliwowe elektrownie i elektrociepłownie gazowo-parowe. Efektywność energetyczna i ekonomiczna), WNT, Warszawa
7. Bartnik R, Buryn Z, Hnydiuk-Stefan A (2017) Methodology and a continuous time mathematical model for selecting the optimum capacity of a heat accumulator integrated with a CHP plant (in Polish: Ekonomika Energetyki w modelach matematycznych z czasem ciągłym). Wydawnictwo Naukowe PWN, Warszawa
8. Bartnik R, Bartnik B (2014) Economic calculation in the energy sector (in Polish: Rachunek ekonomiczny w energetyce). WNT, Warszawa
9. Bartnik R, Buryn Z, Hnydiuk-Stefan A (2017) Investment strategy in heating and CHP. Mathematical Models, Springer, London
10. Bartnik R, Bartnik B, Hnydiuk-Stefan A (2016) Optimum investment strategy in the power industry. Mathematical Models, Springer, New York

Chapter 4
Thermodynamic and Economic Analysis of Trigeneration System with a Hierarchical Gas-Gas Engine for Production of Electricity, Heat and Cold

4.1 Introduction

Chapter 3 includes the analysis of energy and economic effectiveness of both operation options of the innovative, hierarchical two-cycle *gas-gas* engine using combined two clockwise *Joule-Brayton* cycles, high-temperature *Joule-Brayton* gas turbine cycle and low-temperature *Joule-Brayton* cycle of the turboexpander. The cycles are combined by the *exhaust gas-air* system in the air heater *N*—Fig. 4.1a, b, c. In one of the options, only electricity is generated, while in another one electricity and heat is produced in combination. The energy efficiencies of both options are lower than the efficiency of the two-cycle hierarchical *gas-steam* engine that uses the gas turbine clockwise *Joule-Brayton* cycle operating at a high temperature range and combined steam turbine clockwise *Clausius-Rankine* cycle, operating at a low temperature range—Fig. 1.1.

This chapter presents the results of energy and economic effectiveness of the above-mentioned hierarchical *gas-gas* engine, but the trigeneration system was subject to analysis, the one that produces electricity, heat and cooling in the cogeneration system—Fig. 4.1a, b, c. This solution significantly increases the energy efficiency of the *gas-gas* system as compared to the system without cooling production.

Figure 4.1a presents the compressor chiller installed driven by the electric motor, while Fig. 4.1b presents the compressor chiller driven by mechanical work of the rotating machine shaft, gas turbine and turboexpander. The ease of cooling power adjustment is rather a reason that encourages us to choose the electric motor.

Figure 4.1c presents the schematic diagram of the thermal chiller driven with the lowtemperature enthalpy of exhaust gases delivered from the hierarchical gas-gas engine.

Thanks to this the revenue from selling electricity is not "depleted" for the electricity generated in the system as compared to systems Fig. 4.1a, b. Electricity, as already mentioned above, is used to drive compressor chillers (mechanic work—Fig. 4.1b) that is generated in the system, which reduces the revenue from its sales. Thanks to this compressor chillers have the advantage of high cooling power, that is

R. Bartnik and T. W. Kowalczyk, *Hierarchical Gas-Gas Systems*,
Power Systems, https://doi.org/10.1007/978-3-030-69205-6_4

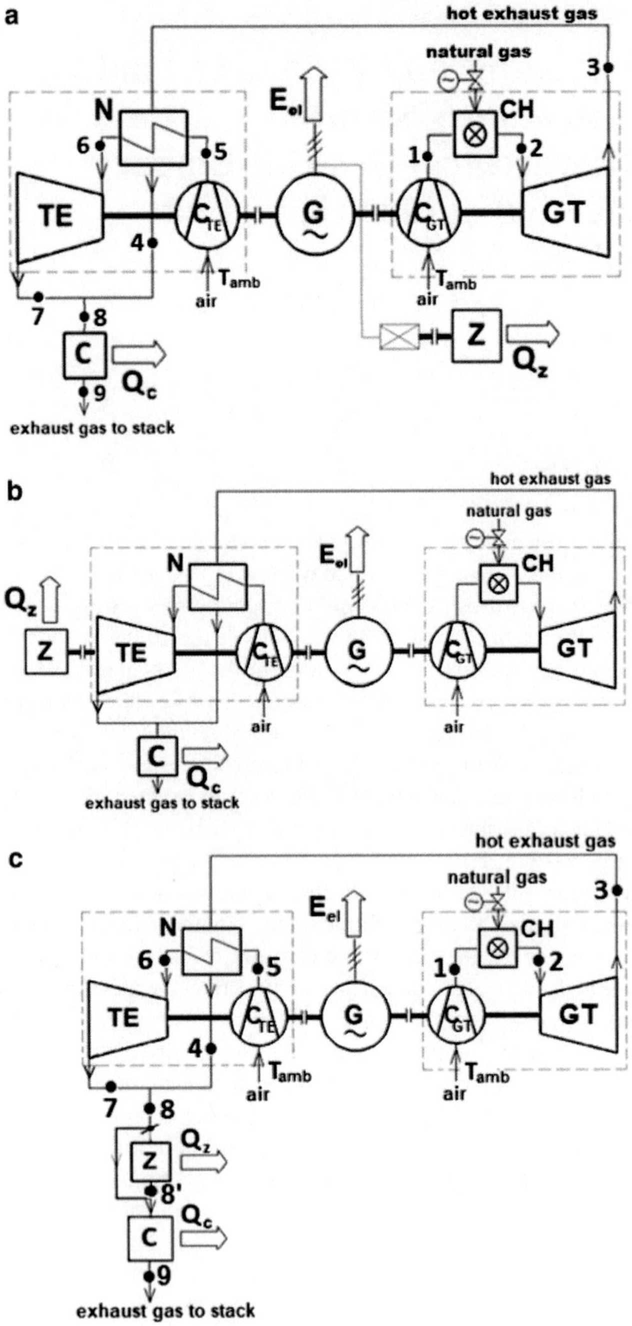
a
hot exhaust gas
natural gas
3
E_el
N
CH
6
5
1
2
TE
C_TE
G
C_GT
GT
4
T_amb
air
T_amb
air
7
8
C
Z
Q_z
9
Q_c
exhaust gas to stack
b
hot exhaust gas
natural gas
E_el
N
CH
Q_z
Z
TE
C_TE
G
C_GT
GT
air
air
C
Q_c
exhaust gas to stack
c
hot exhaust gas
natural gas
3
E_el
N
CH
6
5
1
2
TE
C_TE
G
C_GT
GT
4
T_amb
air
T_amb
air
7
8
Z
Q_z
8'
Q_c
C
9
exhaust gas to stack

◂**Fig. 4.1** **a** The schematic diagram of the system with the compressor chiller driven by the electric motor **b** The schematic diagram of the system with compressor chiller placed on a common shaft with the gas turbine and turboexpander **c** the schematic diagram of the thermal chiller driven with the low-temperature enthalpy of exhaust gases delivered from the hierarchical *gas-gas* engine. **c** The schematic diagram of the gas-gas trigeneration system with the gas turbine and turboexpander in a single-shaft configuration used for combined generation of electricity, heat and cooling in the thermal chiller (G—electric generator, CH—gas turbine combustion chamber, N—air heater, the heater is a device that combines the gas turbine cycle with the turboexpander cycle, C_{TE}—low-pressure compressor, C_{GT}—gas turbine high-pressure compressor, GT—turbine part of the gas turbine, TE—turboexpander, C—heating exchanger, Z—thermal chiller, 1 ÷ 9—system points)

high cooling power sales, while in combination with the thermal chiller its production is relatively low. It results from a limited available exhaust gases range of temperatures used to drive the chiller. Also relatively low is the production of heating heat, because the enthalpy of exhaust gases is used not only to produce it, but, as already mentioned, to provide cooling—Fig. 4.1c.

In the trigeneration system under analysis, the same as in cogeneration systems (Chap. 3), it is advantageous to install the turboexpander and low-pressure compressor in a single housing and on the common shaft, which is symbolically illustrated in Fig. 4.1a, b, c by using squares drawn with a dotted line, the same way as it is in real for the turbine part and the compressor of the high-pressure gas turbine. This solution reduces investment outlays on the *gas-gas* engine.

By analysing the operation of the hierarchical, trigeneration *gas-gas* system it is justified to find an answer to the question: to what extent the use of heat recovery in the turboexpander—Fig. 4.2—as well as in the gas turbine will increase its power: electric, heating and cooling? A significant increase in any medium would translate in a higher economic viability of system operation. Here, it should be stressed that using regeneration does not increase the chemical energy stream $\dot{E}_{ch}$ of gas used to drive the gas turbine, hence does not increase the cost of fuel. Because in the hierarchical system fuel is supplied only to the cycle operating at the highest temperature ranges. Each of subsequent cycles in the hierarchical system operating at lower and lower temperature ranges—the driving heat supplied to them is the heat emitted from cycles located immediately above them in the hierarchy.

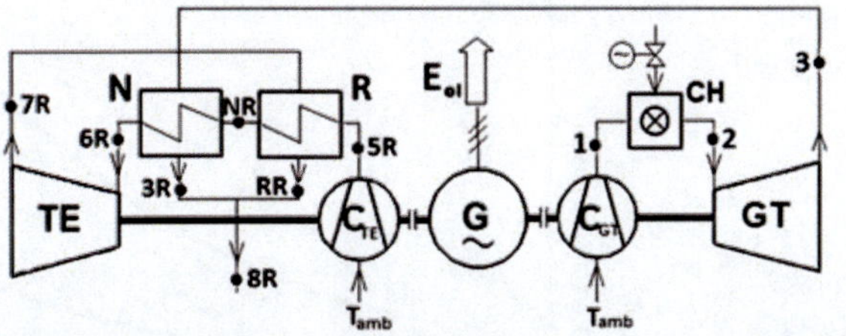

Fig. 4.2 The schematic diagram of the hierarchical gas-gas engine with the turboexpander operating with heat recovery (R—regenerative heat exchanger)

4.2 Thermodynamic and Economic Analysis of a Gas-Gas System for Combined Electricity, Heat and Cooling Generation

The thermodynamic analysis allows us to find an opportunity to improve technology processes and construction solutions for machines and devices. In a market economy it is the economic criterion, profit and its maximisation that decide whether a specific technical solution should be used, it is the economic viability that makes the basis for making a decision. This is because the economic criterion is superior to the technical criterion. However, it should be emphasised that the economic analysis is possible only after a previous thermodynamic analysis. Its results make up input values for the economic analysis. As an example, the Fig. 4.3 presents a graphical interpretation of the thermodynamic model of the Joule-Brayton cycle in the temperature-entropy diagram.

4.2.1 Analysis of a Compressor Chiller System

The input values for the economic analysis of the trigeneration system—Fig. 4.1a, b, c—the same, as it is in the economic analysis of the gas-gas system in Chap. 3 are: gas turbine power, N^{TG}turboexpander power N^{TE}, thermal power of the $\dot{Q}_c$ heating exchanger, cooling power $\dot{Q}_z$ of the compressor chiller, compressor or thermal and the chemical energy stream of gas $\dot{E}_{ch}$ used in the gas turbine. The power of the gas turbine N^{TG} determines all the remaining quantities: $N^{TE} = f(N^{TG})$, $\dot{Q}_c = f(N^{TG})$, $\dot{Q}_z = f(N^{TG})$ and, which is obvious, $\dot{E}_{ch} = f(N^{TG})$. Hence, all the listed power values of the trigeneration system depend on the power of the gas turbine N^{TG} and the temperature of exhaust gases delivered to it T_2. Another extremely important issue is to develop a mathematical model of economic viability of the trigeneration system operation in such a way to use them to generalise the considerations, i.e. to use the calculation results for any gas turbine power N^{TG},

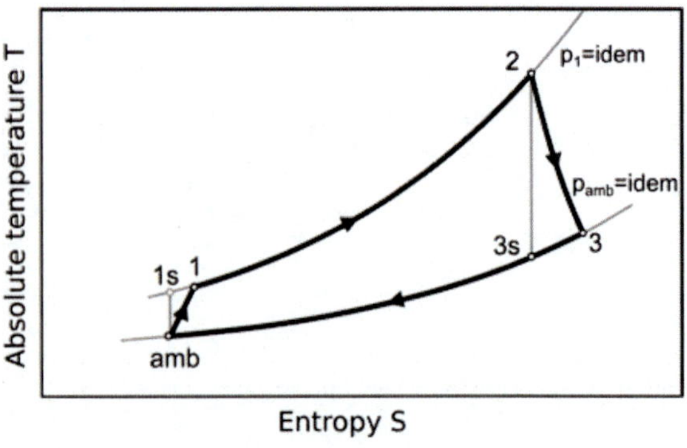

Fig. 4.3 *Joule-Brayton* cycle of the gas turbine

so any power of the trigeneration system. To achieve, the same as in Sect. 3.3, the model was provided only with quantities without specific values (dimensionless): N^{TE}/N^{TG}, $\dot{Q}_c/N^{TG}$, $\dot{Q}_z/N^{TG}$, $\dot{E}_{ch}/N^{TG}$. Because only this “dimensionless” approach allows us to generalise the considerations, i.e. evaluate the economic viability of the trigeneration system with any power: electrical, thermal and cooling. Therefore, the results presented in the chapter are universal and they refer to any system power. The power values are determined only for the gas turbine power N^{TG} and temperature T_2 of exhaust gas delivered to it from CH combustion chamber of a gas turbine. Temperature T_2 is the basic value that characterises turbine operation and is always provided in catalogues by manufacturers. As a last resort, to find out the unit cost of cooling generation in the presented in the monograph trigeneration systems, it is suffice to know the temperature T_2 of exhaust gases delivered to the gas turbine and the current prices of gas, electricity and heat—Figs. 4.4, 4.5, 4.6, 4.7, 4.8, 4.9, 4.12, 4.13, 4.14, 4.15, 4.16 and 4.17.

The fundamental values that affect the viability of trigeneration *gas-gas* systems include energy efficiency of the gas turbine η_{TG} and turboexpander η_{TE} using the *Joule-Brayton* cycle. They can be determined by using energy balances.

From the energy balance in the case of *Joule-Brayton* cycle of the gas turbine we obtain [1]:

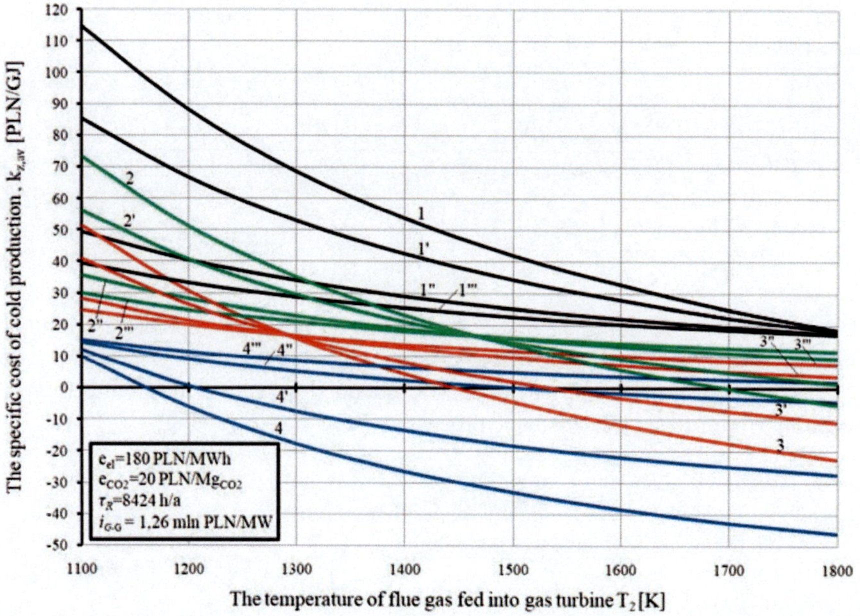

Fig. 4.4 The values of unit cooling production costs $k_{z,av}$ in the gas-gas system with the turboexpander with no heat recovery with the compressor chiller as the temperature function of inlet exhaust gases to the gas turbine T_2

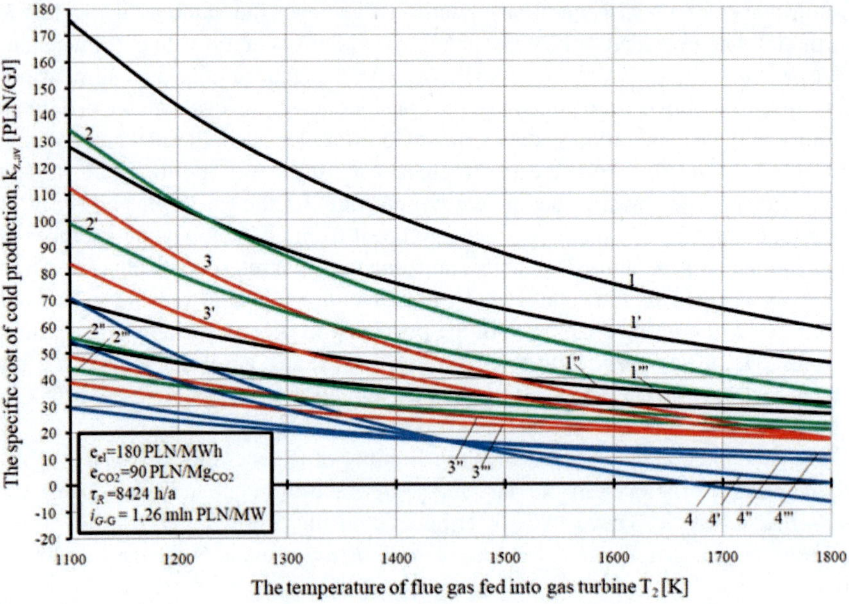

Fig. 4.5 The values of unit cooling production costs $k_{z,\,av}$ in the gas-gas system with the turboexpander with no heat recovery with the compressor chiller as the temperature function of inlet exhaust gases to the gas turbine T_2

$$\eta_{GT} = \frac{N^{GT}}{\dot{E}_{ch}} = \frac{N_{i\ \exp} - N_{i\ C}}{\dot{E}_{ch}} = \frac{\eta_m(T_2 - T_3) - \frac{1}{\eta_m}(T_1 - T_{amb})}{T_2 - T_1} = \\ = \frac{\eta_m^2 \eta_i^C \eta_i^{GT} T_2\left(1 - \frac{1}{z_{GT}}\right) - T_{amb}(z_{GT} - 1)}{\eta_m \eta_i^C (T_2 - T_{amb}) - \eta_m T_{amb}(z_{GT} - 1)} \rightarrow \max \tag{4.1}$$

Efficiency η_{GT} is higher the higher is temperature T_2. This temperature is limited by heat resistance of the turbine blade system.

The final form of formula (4.1) can be obtained by substituting temperatures resulting from irreversible adiabatic processes amb-1 and 2–3—Fig. 4.3:

$$T_1 = T_{amb} + \frac{1}{\eta_i^C}(T_{1s} - T_{amb}) \tag{4.2}$$

$$T_3 = T_2 - \eta_i^{GT}(T_2 - T_{3s}) \tag{4.3}$$

where:

$$z_{GT} = \frac{T_{1s}}{T_{amb}} = \frac{T_2}{T_{3s}} = \left(\frac{p_1}{p_{amb}}\right)^{\frac{\kappa-1}{\kappa}} \tag{4.4}$$

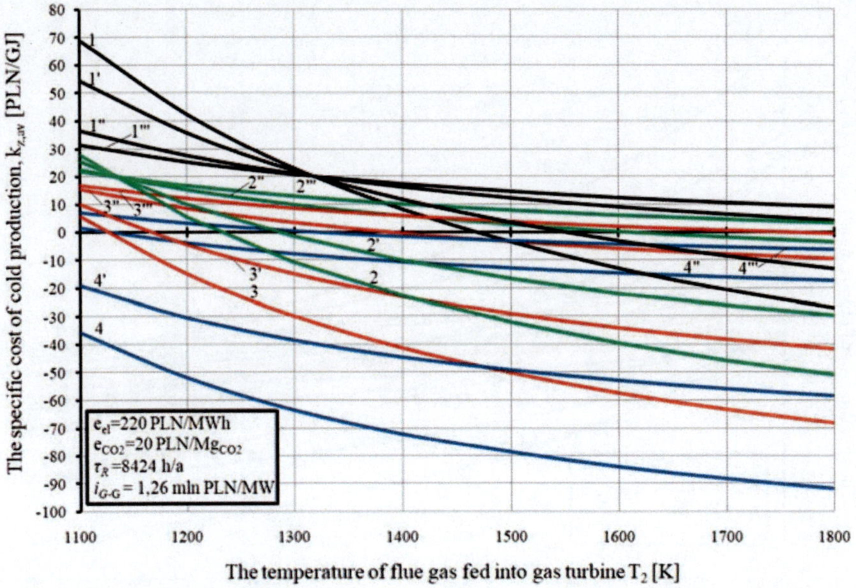

Fig. 4.6 The values of unit cooling production costs $k_{z,\,av}$ in the gas-gas system with the turboexpander with no heat recovery with the compressor chiller as the temperature function of inlet exhaust gases to the gas turbine T_2

From formula (4.1) we can determine the optimum ratio of pressures z_{GT}^{opt}, for which the cycle assumes the maximum efficiency $\eta_{GT}^{\max}$, i.e. maximum power (the same proceedings as for calculating the efficiency of $\eta_{TE}^{\max}$ the turboexpander). The ratio z_{GT}^{opt} is the function of temperatures T_{amb}, T_2 and mechanical efficiencies η_m of the compressor and turbine (it was assumed that mechanical efficiencies of the compressor and turbine are the same) as well as internal compressor η_i^C and turbine efficiencies η_i^{GT}:

$$z_{TG}^{opt}\left(T_{amb}, T_2, \eta_m, \eta_i^C, \eta_i^{GT}\right) = \left(\frac{p_1}{p_{amb}}\right)_{opt}^{\frac{\kappa-1}{\kappa}} \tag{4.5}$$

where:

κ	exponent of isentrope of the heat transfer medium (for calculations it was assumed $\kappa = 1.4$),
$p_1 = p_2, p_{amb}$	pressures of the heat transfer medium during heat absorption and emission (for calculations it was assumed $p_{amb} = 0.1$ MPa).

For thermodynamic calculations it was assumed that ambient temperature is T_{amb} = 288 K, while mechanical efficiencies of the compressor and gas turbine are equal

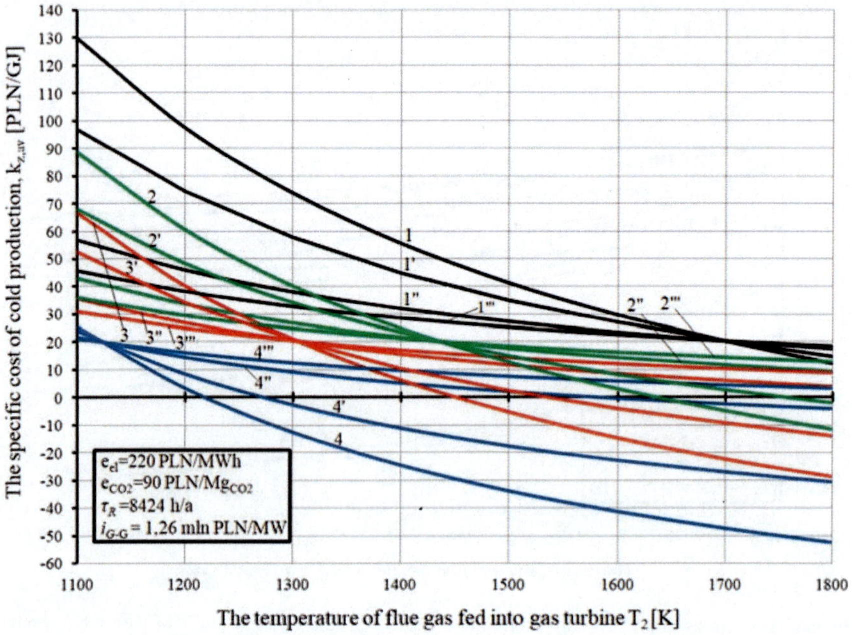

Fig. 4.7 The values of unit cooling production costs $k_{z,av}$ in the gas-gas system with the turboexpander with no heat recovery with the compressor chiller as the temperature function of inlet exhaust gases to the gas turbine T_2

and amount to $\eta_m^S = \eta_m^{GT} = \eta_m = 0.97$ and their internal efficiencies equal $\eta_i^{GT} = 0.87$, $\eta_i^C = 0.85$.

The optimum value z_{GT}^{opt} derives from the condition:

$$\frac{d\eta_{GT}}{dz_{GT}} = 0 \tag{4.6}$$

After differentiating the equation (4.1) and using the condition (4.6) we obtain:

$$(bc - ad)(z_{GT}^{opt})^2 - 2bcz_{GT}^{opt} + b(c + d) = 0 \tag{4.7}$$

therefore

$$z_{GT}^{opt} = \frac{bc - \sqrt{bd(ad + ac - bc)}}{(bc - ad)} \tag{4.8}$$

where:

$$a = T_{amb},\ b = \eta_m^2 \eta_i^C \eta_i^{GT} T_2,\ c = \eta_m T_{amb},\ d = \eta_m \eta_i^C (T_2 - T_{amb}). \tag{4.9}$$

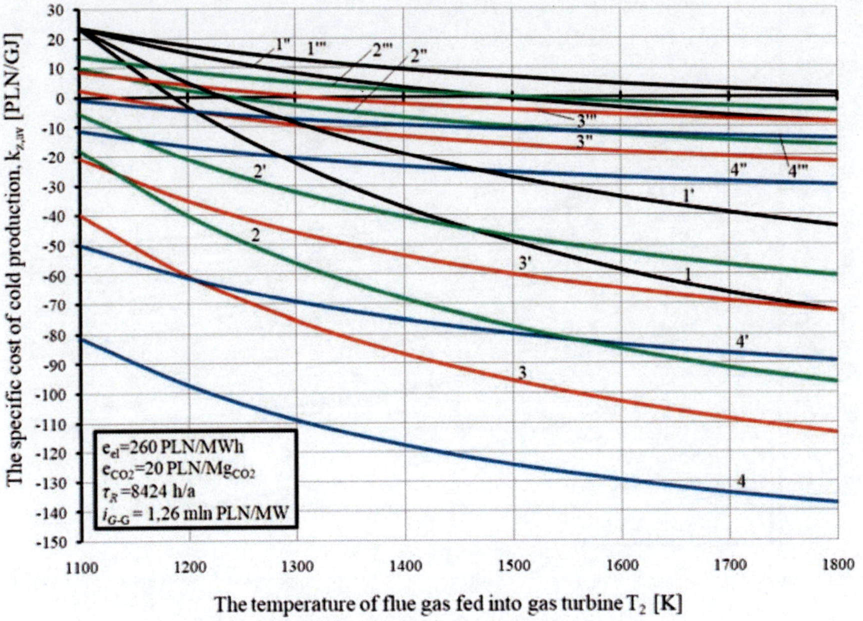

Fig. 4.8 The values of unit cooling production costs $k_{z,av}$ in the gas-gas system with the turboexpander with no heat recovery with the compressor chiller as the temperature function of inlet exhaust gases to the gas turbine T_2

The second root of the equation (4.7) is unreal, because the temperature T_1 would be higher than temperature T_2.

The maximum value $\eta_{GT}^{\max}$ can be calculated from formula (4.1) substituting z_{GT} the value z_{GT}^{opt}. As a result, for $\eta_{GT}^{\max}$ and specific value $\dot{E}_{ch}$ we calculate the maximum power of the gas turbine used in the economic analysis. In the available gas turbines z_{GT} assume obviously values z_{GT}^{opt}.

Temperatures of the heat transfer medium behind the compressor T_1 and behind the turbine T_3 are determined, as already mentioned, by using irreversible adiabatic processes:

$$T_1 = T_{amb}\left[1 + \frac{1}{\eta_i^C}\left(z_{GT}^{opt} - 1\right)\right] \tag{4.10}$$

$$T_3 = T_2\left[1 - \eta_i^{GT}\left(1 - \frac{1}{z_{GT}^{opt}}\right)\right] \tag{4.11}$$

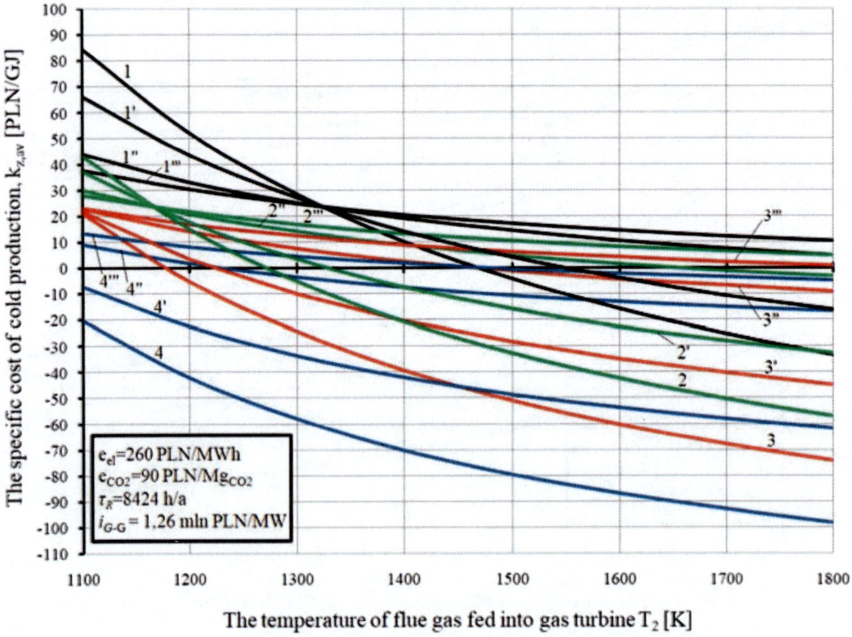

Fig. 4.9 The values of unit cooling production costs $k_{z,av}$ in the gas-gas system with the turboexpander with no heat recovery with the compressor chiller as the temperature function of inlet exhaust gases to the gas turbine T_2

As already mentioned, the same way we can determine z_{TE}^{opt} and $\eta_{TE}^{\max}$, but, as temperatures T_1, T_2 and T_3 in formulas (4.1), (4.5), (4.9), (4.10) and (4.11) we have to substitute temperatures respectively T_5, T_6 and T_7, while pressure p_1 in formula (4.5) should be substituted with the pressure p_5. Resulting from value $\eta_{GT}^{\max}$, $\eta_{TE}^{\max}$ maximum power N^{GT}, N^{TE} are input values for economic calculations (4.12). The results of thermodynamic calculations for the gas turbine set and turboexpander are presented in Figs. 4.21, 4.22 and 4.23.

As already mentioned, the power values N^{GT}, N^{TE} determined in the thermodynamic analysis are input values for the economic analysis. Its purpose is to determine a unit cost for cooling production. It can be determined by using formula for a discounted *NPV* profit obtained within *T* years of the trigeneration gas-gas system for combined electricity, heat and cooling production:

$$NPV = \Bigg\{\Big[\Big(N^{GT} + N^{TE}\Big)\eta_G \tau_A - N_z\tau_z\Big](1-\varepsilon_{el})\frac{e_{el}^{t=0}}{a_{el}-r}[e^{(a_{el}-r)T} - 1] + Q_{A\,c}\frac{e_c^{t=0}}{a_c - r}[e^{(a_c - r)T} - 1] +$$

$$+ Q_{A\,z}\frac{e_z^{t=0}}{a_z - r}[e^{(a_z - r)T} - 1] - \dot{E}_{ch}\tau_A\Bigg\{\frac{e_{fuel}^{t=0}}{a_{fuel} - r}[e^{(fuel - r)T} - 1] +$$

$$+ \frac{\rho_{CO_2} p_{CO_2}^{t=0}}{a_{CO_2} - r}[e^{(a_{CO_2} - r)T} - 1] + \frac{\rho_{CO} p_{CO}^{t=0}}{a_{CO} - r}[e^{(a_{CO} - r)T} - 1] +$$

$$+ \frac{\rho_{NO_X} p_{NO_X}^{t=0}}{a_{NO_X} - r}[e^{(a_{NO_X} - r)T} - 1] + \frac{\rho_{SO_2} p_{SO_2}^{t=0}}{a_{SO_2} - r}[e^{(a_{SO_2} - r)T} - 1] +$$

$$+ \frac{\rho_{dust} p_{dust}^{t=0}}{a_{dust} - r}[e^{(a_{dust} - r)T} - 1] + \frac{\rho_{CO_2} e_{CO_2}^{t=0}}{b_{CO_2} - r}[e^{(b_{CO_2} - r)T} - 1]\Bigg\} +$$

$$-(1 + x_{sal,t,ins})(J_{G-G} + J_z)(1 - e^{-rT})\frac{\delta_{serv}}{r} - z(J_{G-G} + J_z)\Big(\frac{1 - e^{-rT}}{T} + 1\Big)\Bigg\}(1 - p) \tag{4.12}$$

wherein the annual heat and cooling production are expressed by the following formulas:

$$Q_{Ac} = \frac{\big(N^{GT} + N^{TE}\big)\eta_G \tau_A}{\sigma_A} \tag{4.13}$$

$$Q_{A\,z} = \dot{Q}_z \tau_z = N_z \varepsilon_z \tau_z \tag{4.14}$$

while compressor chiller driving power with dependency:

$$N_2 = x\big(N^{GT} + N^{GT}\big)\eta_G;\ x \in \langle 0; 1\rangle. \tag{4.15}$$

where:

$a_{el}, a_c, a_z, a_{fuel}, a_{CO_2}, a_{SO_2}, a_{NO_X}, a_{dust}, b_{CO_2}$	exponents illustrating changes in time for electricity, heat, cooling, fuel, environmental fees, CO_2 emission permissions,
$e_{el}^{t=0}, e_{fuel}^{t=0}, e_c^{t=0}, e_z^{t=0}, e_{CO_2}^{t=0}, p_{CO_2}^{t=0}$ etc.	initial values of prices of electricity, fuel, heat, cooling, CO_2 emission permissions, etc.
$\dot{E}_{ch}$	fuel chemical energy stream burnt in the gas turbine ($\dot{E}_{ch} = N^{GT}/\eta_{GT}^{\max}$), heat, cooling and the purchase of CO_2 environmental fees,
J_{G-G}, J_z	investment outlays on the gas-gas system and chiller,
$p_{co_2}, p_{co}, p_{NO_x}, p_{SO_2}, p_{dust}$	unit rate per CO_2, CO, NO_x, SO_2, dust,
r	investment capital interest rate,

$x_{sal,t,ins}$ coefficient taking into account the cost of pays, taxes, insurance policies, etc. (in practice the value $x_{sal,t,ins}$ is approx. 0.25),

z discounting coefficient (freezing ratio) of investment capital J at the moment of investment end,

δ_{serv} annual fixed cost rate depending on investment outlays (costs of maintenance, repairs of devices).

$\varepsilon_z = \dot{Q}_z / N_z$ thermal efficiency of the chiller (for calculations the steam compressor chiller was assumed using the *Linde* cycle; ammonia is the heat transfer medium; for calculations it was assumed that $\varepsilon_z = 3.2$),

ε_{el} index of own electrical demand of the system,

η_G electric generator efficiency,

$\rho_{CO_2}, \rho_{CO}, \rho_{NO_x}, \rho_{SO_2}, \rho_{py}$ CO_2, CO, NO_x, SO_2, dust emission per fuel chemical energy unit,

$\sigma_A = E_{el,A} / Q_{Ac}$ ratio of the annual electricity production to the annual heat production,

τ_A the annual operating time for the trigeneration system (in hours) (for calculations it was assumed that its value was 8,424 h; a two-week long system downtime per year was assumed for maintenance and repairs),

τ_z the annual operating time for the chiller (in hours).

For conditions $NPV = 0$ and $a_z = 0$ we determine an average unit cooling production cost in the period of T years:

$$
\begin{aligned}
k_{z\,av} = {} & \frac{r\tau_A}{x\left(1+\frac{NTE}{NGT}\right)\eta_G \varepsilon_z \tau_z (1-e^{-rT})\eta_{GT}^{\max}} \left\{ \frac{e_{fuel}^{t=0}}{a_{fuel}-r}[e^{(a_{fuel}-r)T}-1] + \right. \\
& + \frac{\rho_{CO_2} p_{CO_2}^{t=0}}{a_{CO_2}-r}[e^{(a_{CO_2}-r)T}-1] + \frac{\rho_{CO} p_{CO}^{t=0}}{a_{CO}-r}[e^{(a_{CO}-r)T}-1] + \\
& + \frac{\rho_{NO_X} p_{NO_X}^{t=0}}{a_{NO_X}-r}[e^{(a_{NO_X}-r)T}-1] + \frac{\rho_{SO_2} p_{SO_2}^{t=0}}{a_{SO_2}-r}[e^{(a_{SO_2}-r)T}-1] + \\
& \left. + \frac{\rho_{dust} p_{dust}^{t=0}}{a_{dust}-r}[e^{(a_{dust}-r)T}-1] + \frac{\rho_{CO_2} e_{CO_2}^{t=0}}{b_{CO_2}-r}[e^{(b_{CO_2}-r)T}-1] \right\} + \\
& + (1+x_{sal,t,ins})\frac{(i_{G-G}+xi_z)}{x\,\varepsilon_z \tau_z}\delta_{serv} + \frac{rz(i_{G-G}+xi_z)}{x\,\varepsilon_z \tau_z (1-e^{-rT})}\left(\frac{1-e^{-rT}}{T}+1\right) + \\
& - \frac{r(\tau_A - x\,\tau_z)(1-\varepsilon_{el})}{x\,\varepsilon_z \tau_z (1-e^{-rT})}\frac{e_{el}^{t=0}}{a_{el}-r}[e^{(a_{el}-r)T}-1] - \frac{r\tau_A}{x\,\varepsilon_z \tau_z \sigma_A (1-e^{-rT})}\frac{e_c^{t=0}}{a_c-r}[e^{(a_c-r)T}-1]
\end{aligned}
\tag{4.16}
$$

where:

i_{G-G} unit (per electric power unit) investment outlay on the gas-gas engine, $i_{G-G} = J_{G-G} / \left[\left(N^{TG} + N^{TE} \right) \eta_G \right]$,

i_z unit (per electric power unit) investment outlay on the chiller, $i_z = J_z / N_z$ (for calculations two options of outlay amounts were assumed; $i_z = 0, 5 i_{G-G}$, $i_z = 0, 7 i_{G-G}$).

The revenues from selling electricity and heating (formulas (4.12) and (4.16)) are avoided costs of cooling production,

One of the values that apart from investment outlays, gas turbine set energy efficiency, power and operating time of the chiller, prices of electricity, fuel and heat have a significant impact on the unit cooling production cost $k_{z\ av}$ is the price of carbon dioxide emission permissions e_{CO_2}. It is subject to very high fluctuations, even week to week, as it is frequently a speculative price. For comparative purposes, its value was assumed in a multi-variant way. In one approach it amounts to the so-called reference price $e_{CO_2}^{ref} = 20.38$ EUR/Mg_{CO2} (about 90 PLN/Mg_{CO2}) that according to Directive 2003/87/EC and later 2009/29/EC imposes the trading of CO_2 *EU ETS* emissions (European Union Emission Trading Scheme), and in the other approach – $e_{CO_2} = 20$ PLN/Mg_{CO2}.

Formula (4.16) can be used to determine also unit cooling production costs for all the three variants of systems operating with heat recovery, i.e. for systems with the regeneration heat exchanger used: (1) only in the gas turbine, (2) only in the turboexpander and (3) for the system, where regeneration heat exchangers are simultaneously used in the gas turbine and turboexpander. They are different in formula (4.16) for individual variants, resulting from the thermodynamic analysis, power ration values N^{TE} / N^{GT}, as well as, different are also, slightly higher, the unit investment outlays.

Figures 4.4, 4.5, 4.6, 4.7, 4.8 and 4.9 present the results of multi-variant calculations of the unit cooling production cost in the system with the compressor chiller. Table 4.1 lists calculation input data for individual curves presented in Figs. 4.4, 4.5, 4.6, 4.7, 4.8 and 4.9.

As shown in Figs. 4.4, 4.5, 4.6, 4.7, 4.8 and 4.9 unit cooling production costs $k_{z,\ av}$ are higher, which is obvious, the lower the prices of electricity, fuel and CO_2 emission permissions and the higher investment outlays. Increasing x at the same time, i.e. increasing the cooling production in the system (formulas (4.14), (4.15)), will raise this cost. This is the case, because then the electricity production decreases, which reduces the revenue from its sales making up the avoided cost of cooling production.

While analysing the system operation it is justified to find an answer to the question: to what extent does the use of heat recovery in the turboexpander, Fig. 4.2, increasing electricity production (Sect. 4.3) affect the unit cooling production cost? Assuming that investment outlays on the system with recovery are the same as outlays on the system without the regenerator, it turns out that reducing the cost is negligible and does not exceed 0.5 PLN/GJ within the whole temperature change range T_2. Therefore, the curves for unit costs presented in Figs. 4.4, 4.5, 4.6, 4.7, 4.8 and 4.9 for the system without regeneration "merge" with the curves for the system with regeneration. So, it can be concluded that the construction of the *gas-gas* system with the turboexpander with heat recovery, when it takes into account the increase

Table 4.1. Input data list

	1	**1'**	**1"**	**1'''**	**2**	**2'**	**2"**	**2'''**
e_c, PLN/GJ	50	50	50	50	70	70	70	70
e_{fuel}, PLN/GJ	24	24	24	24	24	24	24	24
i_z, PLN/MW	$0,5i_{G-G}$	$0,7i_{G-G}$	$0,5i_{G-G}$	$0,7i_{G-G}$	$0,5i_{G-G}$	$0,7i_{G-G}$	$0,5i_{G-G}$	$0,7i_{G-G}$
τ_z, h/a	$0,7\tau_A$	$\tau_z = \tau_A$	$0,7\tau_A$	$\tau_z = \tau_A$	$0,7\tau_A$	$\tau_z = \tau_A$	$0,7\tau_A$	$\tau_z = \tau_A$
x, %	0.1	0.1	0.3	0.3	0.1	0.1	0.3	0.3
	3	**3'**	**3"**	**3'''**	**4**	**4'**	**4"**	**4'''**
e_c, PLN/GJ	50	50	50	50	70	70	70	70
e_{fuel}, PLN/GJ	20	20	20	20	20	20	20	20
i_z, PLN/MW	$0,5i_{G-G}$	$0,7i_{G-G}$	$0,5i_{G-G}$	$0,7i_{G-G}$	$0,5i_{G-G}$	$0,7i_{G-G}$	$0,5i_{G-G}$	$0,7i_{G-G}$
τ_z, h/a	$0,7\tau_A$	$\tau_z = \tau_A$	$0,7\tau_A$	$\tau_z = \tau_A$	$0,7\tau_A$	$\tau_z = \tau_A$	$0,7\tau_A$	$\tau_z = \tau_A$
x, %	0.1	0.1	0.3	0.3	0.1	0.1	0.3	0.3

in outlays in the economic calculation, so when it considers the increase in capital costs (depreciation and financial costs) and operating costs (repairs and maintenance) [2–4], then the increase in the unit cooling production cost in the best option will amount to zero. Therefore, constructing the system with heat recovery is completely economically unviable.

Furthermore, the unit cooling cost analysis presented in Figs. 4.4, 4.5, 4.6, 4.7, 4.8 and 4.9 shows, as indicated above, that it is more profitable to produce only electricity and heat in the system than use it additionally to provide cooling. Consequently, very important is the answer to the question: what should be the price of cooling to make its production profitable?

4.2.1.1 Economic Viability of Using a Compressor Chiller in a Trigeneration System with a Gas-Gas Engine

The prerequisite for the economic viability of using the compressor chiller in the trigeneration system is that *NPV* profit from its operation (4.12) should be higher than the profit from the system operation used (only to generate electricity and heat)—in formulas (4.12), (4.15) zero value should be substituted for quantities N_z and J_z. This prerequisite can be expressed by the following dependency:

$$Q_{A\,z}\frac{e_{z\,\min}^{t=0}}{a_z - r}[e^{(a_z - r)T} - 1] \geq N_z\tau_z(1-\varepsilon_{el})\frac{e_{el}^{t=0}}{a_{el} - r}[e^{(a_{el}-r)T} - 1] + \\ + (1 + x_{sal,t,ins})J_z(1 - e^{-rT})\frac{\delta_{serv}}{r} + zJ_z\left(\frac{1 - e^{-rT}}{T} + 1\right) \tag{4.17}$$

Using Eq. (4.15) we obtain (4.17), the final form of prerequisite average sales price ($a_z = 0$) of cooling in the period of T years:

$$e_{z\,av}^{\min} \geq \frac{(1-\varepsilon_{el})}{\varepsilon_z}e_{el\,av} + (1 + x_{sal,t,ins})\frac{i_z\delta_{serv}}{\varepsilon_z\tau_z} + \frac{rzi_z}{\varepsilon_z\tau_z(1 - e^{-rT})}\left(\frac{1 - e^{-rT}}{T} + 1\right) \tag{4.18}$$

while the average price $e_{el\,av}$ of selling electricity in the period of T years can be expressed by the following formula:

$$e_{el\,av} = \frac{1}{T}\int_0^T e_{el}^{t=0}e^{a_{el}t}dt = \frac{e_{el}^{t=0}}{Ta_{el}}(e^{a_{el}T} - 1) \tag{4.19}$$

However, it should be strongly emphasised that the prerequisite for trigeneration system viability is not to meet the relation (4.18), but it should meet the dependency:

$$e_{el\,av} \geq k_{z\,av} \tag{4.20}$$

i.e. to make the cooling sales price at least not lower than the unit cost of its generation. This is because it might happen that the price $e_{z\,av}^{\min}$ (4.18) is lower than the cost $k_{z\,av}$ (4.16). It takes place for the relatively low electricity cost, generally for high cost $k_{z\,av}$—Fig. 4.10. Therefore, price $e_{z\,av}^{\min}$ as opposed to cost $k_{z\,av}$ is calculated without fuel and environmental fees that equalise in the process of deducting *NPV* profits i.e. when deducting from *NPV* profit from operating the trigeneration system the *NPV* profit obtained for operating the cogeneration system (4.18).

The forms of example curves $e_{z\,av}^{\min}$ and $k_{z\,av}$ are presented in Fig. 4.10. Table 4.2 lists input data assumed for their calculation.

4.2.2 *Analysis of a Thermal Absorption Chiller System*

The advantage of the absorption chiller is the lack of the compressor and not using electricity to power it, but the low-temperature enthalpy of exhaust gases from the *gas-gas* engine—Fig. 4.1c. Thanks to this, as compared to the system with the compressor chiller, the system produces more electricity, the most noble, and also, the most precious and the most expensive energy form, while its sales revenue is the

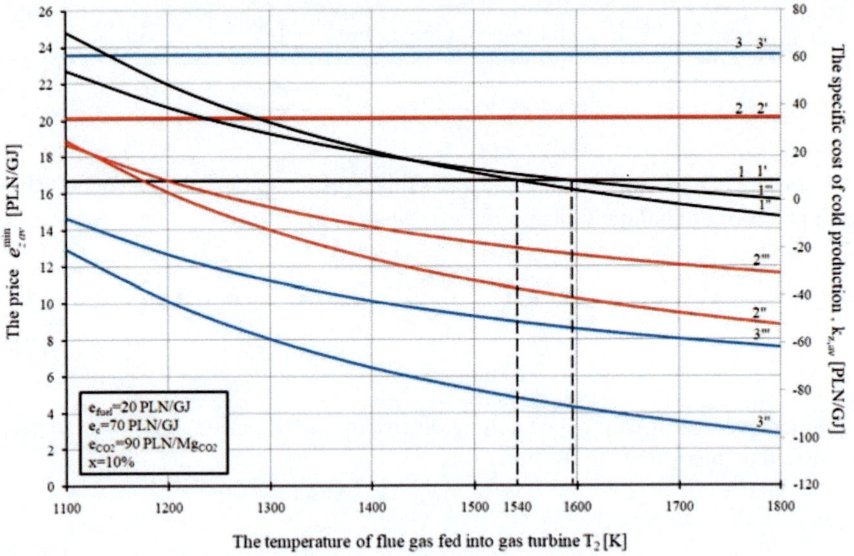

Fig. 4.10 Values $e_{z\,av}^{\min}$, $k_{z\,av}$ as the temperature function of exhaust gases delivered to the gas turbine T_2

Table 4.2. Input data list

e_{el}, PLN/MWh	180	180	220	220	260	260
i_z, PLN/MW	$0,5 i_{G-G}$	$0,7 i_{G-G}$	$0,5 i_{G-G}$	$0,7 i_{G-G}$	$0,5 i_{G-G}$	$0,7 i_{G-G}$
τ_z, h/a	$0,7 \tau_A$	$\tau_z = \tau_A$	$0,7 \tau_A$	$\tau_z = \tau_A$	$0,7 \tau_A$	$\tau_z = \tau_A$
$e_{z\,av}^{\min}$, PLN/GJ	**1**	**1'**	**2**	**2'**	**3**	**3'**
$k_{z,av}$, PLN/GJ	**1''**	**1'''**	**2''**	**2'''**	**3''**	**3'''**

avoided cost of cooling production. The disadvantage of the system with the absorption chiller is its low cooling power. The temperature of exhaust gases constituting the source of driving heat in the desorber (boiler) of the absorption chiller must be higher than the temperature of ammonia solution delivered to it. Temperature amounts to approx. $T_{8'} = 395$ K, while the temperature T_8 of exhaust gases delivered to the desorber ranges from approx. 458 K to approx. 524 K (this temperature level increases with combustion temperature T_2 of gas in the gas turbine combustion chamber CH; value T_8= 458 K corresponds to temperature T_2= 1,100 K, value T_8= 524 K corresponds to temperature T_2= 1,800 K). Therefore, the available range of the temperature of exhaust gases that can be used to drive the chiller is relatively low and ranges from approx. 63 K to approx. 130 K, which translates, as compared to the compressor chiller, to a relatively low heat stream value $\dot{Q}_z$ collected from the chiller chamber. The cooling power $\dot{Q}_z$ of the absorption chiller can be expressed by the following equation:

$$\dot{Q}_z = (\dot{C}_{eg} + \dot{C}_{air})(T_8 - 395)\varepsilon_{z\,abs} \tag{4.21}$$

While the annual cooling production can be expressed by the following dependency:

$$Q_{A\,z} = \dot{Q}_z \tau_z = (\dot{C}_{eg} + \dot{C}_{air})(T_8 - 395)\varepsilon_{z\,abs}\tau_z \tag{4.22}$$

wherein the energy balances of gas turbine and turboexpander cycles can be used to determine thermal capacity streams $\dot{C}_{eg}$, $\dot{C}_{air}$ of *Joule-Brayton* cycles of the gas turbine *TG* and turboexpander *TE*:

$$N^{GT} = \dot{C}_{eg}\left[\eta_m(T_2 - T_3) - \frac{(T_1 - T_{amb})}{\eta_m}\right] \tag{4.23}$$

$$N^{TE} = \dot{C}_{air}\left[\eta_m(T_6 - T_7) - \frac{(T_5 - T_{amb})}{\eta_m}\right] \tag{4.24}$$

where:

N^{GT}, N^{TE}	power levels of the gas turbine *TG* and the turboexpander *TE*, Fig. 4.1c,
T_1	temperature of the heat transfer medium behind the compressor C_{TG}, Fig. 4.1c,
T_2	temperature of the inlet heat transfer medium delivered to *GT*, Fig. 4.1c,
T_3	temperature of the outlet heat transfer medium removed from *GT*, Fig. 4.1c,
T_5	temperature of the heat transfer medium behind the compressor C_{TE}, Fig. 4.1c,
T_6	temperature of the inlet heat transfer medium delivered to *TE*, Fig. 4.1c,
T_7	temperature of the outlet heat transfer medium removed from the turboexpander *TE*, Fig. 4.1c,
T_8	temperature of the driving inlet heating medium delivered to the chiller, Fig. 4.1c,
$\varepsilon_{z\,abs}$	thermal efficiency of the cooling process of the thermal absorption chiller; for calculations it was assumed $\varepsilon_{z\,abs} = 0,37$,
τ_z	the annual operating time for the chiller (in hours).

Figure 4.11 presents the results of thermodynamic calculations of the ratio of the annual heat production to the annual electricity production $Q_{Ac}/E_{el,A}$ and the value of the ratio of the annual cooling production to the annual electricity production $Q_{Az}/E_{el,A}$ in the system with the thermal chiller. Here, it should be stressed that the ratio $Q_{Ac}/E_{el,A}$ in the system with the compressor chiller is about 3 times higher, because the enthalpy of outlet exhaust gases from the engine is solely used to produce

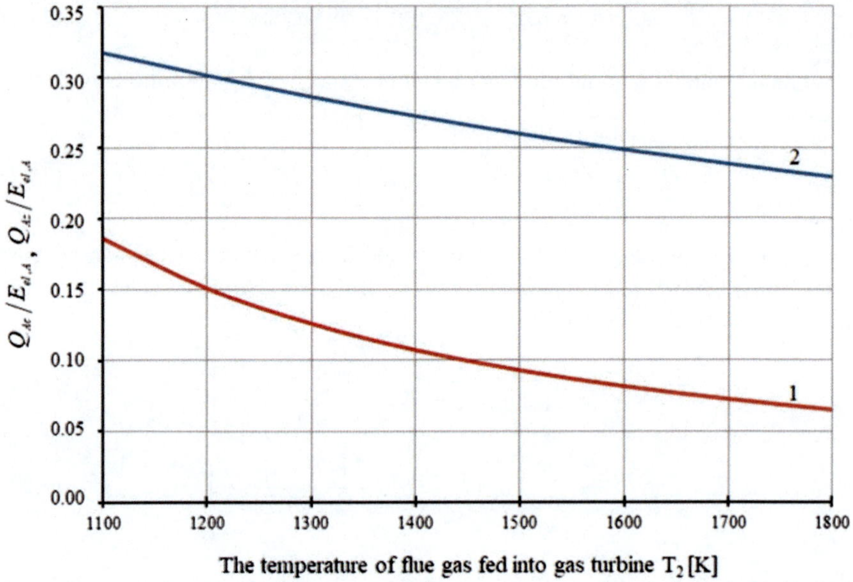

Fig. 4.11 The ratio values of the annual heat and cooling production to the annual electricity production in the *gas-gas* system with the thermal absorption chiller as the function of the temperature of inlet exhaust gases delivered to the gas turbine T_2 (1:$Q_{Ac}/E_{el,A}$; 2: $Q_{Az}/E_{el,A}$)

heat – Fig. 4.1a, b. This ratio, similarly to the system with the thermal chiller, depends also on the temperature T_2 and within the range $T_2 \in \langle 1100;\ 1800\ \text{K}\rangle$ decreases from values 1/2.16 to 1/3.77. Also the cooling power of the system with the compressor chiller, which depends on value x (4.15), is significantly higher as a result of driving the chiller by means of electricity (4.14). In addition, the thermal efficiency equals $\varepsilon_z = 3.2$, when thermal $\varepsilon_{z\ abs} = 0, 37$. When $x \approx 0.1$, the cooling power of the compressor chiller equals the power of the thermal chiller. As a result, the revenue from selling heat in the system with the compressor chiller—Figs. 4.1a, b—is higher by about 3 times than the revenue in the system with the thermal chiller. By contrast, in the system with the compressor chiller the revenue from selling electricity decreases due to its use to drive the chiller—the more, the higher is the value x.

The total discounted *NPV* profit resulting from the operation of the trigeneration system with the thermal absorption chiller can be presented by the following dependency:

$$
\begin{aligned}
NPV = &\left\{\left[\left(N^{GT}+N^{TE}\right)\eta_G\tau_A\right](1-\varepsilon_{el})\frac{e_{el}^{t=0}}{a_{el}-r}\left[e^{(a_{el}-r)T}-1\right]+Q_{A\,c}\frac{e_c^{t=0}}{a_c-r}\left[e^{(a_c-r)T}-1\right]+\right.\\
&+Q_{A\,z}\frac{e_z^{t=0}}{a_z-r}\left[e^{(a_z-r)T}-1\right]-\dot{E}_{ch}\tau_A\left\{\frac{e_{fuel}^{t=0}}{a_{fuel}-r}\left[e^{(a_{fuel}-r)T}-1\right]+\right.\\
&+\frac{\rho_{CO_2}p_{CO_2}^{t=0}}{a_{CO_2}-r}\left[e^{(a_{CO_2}-r)T}-1\right]+\frac{\rho_{CO}p_{CO}^{t=0}}{a_{CO}-r}\left[e^{(a_{CO}-r)T}-1\right]+\\
&+\frac{\rho_{NO_X}p_{NO_X}^{t=0}}{a_{NO_X}-r}\left[e^{(a_{NO_X}-r)T}-1\right]+\frac{\rho_{SO_2}p_{SO_2}^{t=0}}{a_{SO_2}-r}\left[e^{(a_{SO_2}-r)T}-1\right]+\\
&+\left.\frac{\rho_{dust}p_{dust}^{t=0}}{a_{dust}-r}\left[e^{(a_{dust}-r)T}-1\right]+\frac{\rho_{CO_2}e_{CO_2}^{t=0}}{b_{CO_2}-r}\left[e^{(b_{CO_2}-r)T}-1\right]\right\}+\\
&\left.-(1+x_{sal,t,ins})(J_{G-G}+J_Z)(1-e^{-rT})\frac{\delta_{serv}}{r}-z(J_{G-G}+J_Z)\left(\frac{1-e^{-rT}}{T}+1\right)\right\}(1-p)
\end{aligned}
\tag{4.25}
$$

while the unit cooling production cost taking that $NPV = 0$ and $a_z = 0$ can be expressed by the following formula:

$$
\begin{aligned}
k_{z\,av} = &\frac{r\tau_A}{A\eta_{GT}^{\max}\varepsilon_{z\,abs}\tau_z(1-e^{-rT})}\left\{\frac{e_{fuel}^{t=0}}{a_{fuel}-r}\left[e^{(a_{fuel}-r)T}-1\right]+\frac{\rho_{CO_2}p_{CO_2}^{t=0}}{a_{CO_2}-r}\left[e^{(a_{CO_2}-r)T}-1\right]+\right.\\
&+\frac{\rho_{CO}p_{CO}^{t=0}}{a_{CO}-r}\left[e^{(a_{CO}-r)T}-1\right]+\frac{\rho_{NO_X}p_{NO_X}^{t=0}}{a_{NO_X}-r}\left[e^{(a_{NO_X}-r)T}-1\right]+\frac{\rho_{SO_2}p_{SO_2}^{t=0}}{a_{SO_2}-r}\left[e^{(a_{SO_2}-r)T}-1\right]+\\
&+\left.\frac{\rho_{dust}p_{dust}^{t=0}}{a_{dust}-r}\left[e^{(a_{dust}-r)T}-1\right]+\frac{\rho_{CO_2}e_{CO_2}^{t=0}}{b_{CO_2}-r}\left[e^{(b_{CO_2}-r)T}-1\right]\right\}+\\
&+\frac{B\eta_G i_{G-G}}{A\varepsilon_{z\,abs}\tau_z}\left[(1+x_{sal.t.ins})\delta_{serv}+\frac{zr}{(1-e^{-rT})}\left(\frac{1-e^{-rT}}{T}+1\right)\right]+\\
&+\frac{i_z}{\tau_z}\left[(1+x_{sal,t,ins})\delta_{serv}+\frac{zr}{(1-e^{-rT})}\left(\frac{1-e^{-rT}}{T}+1\right)\right]+\\
&-\frac{Br\eta_G\tau_A}{A\varepsilon_{z\,abs}\tau_z(1-e^{-rT})}\left\{(1-\varepsilon_{el})\frac{e_{el}^{t=0}}{a_{el}-r}\left[e^{(a_{el}-r)T}-1\right]+\frac{1}{\sigma_A}\frac{e_c^{t=0}}{a_c-r}\left[e^{(a_c-r)T}-1\right]\right\}
\end{aligned}
\tag{4.26}
$$

wherein:

$$
A = \left[\frac{\eta_m}{\eta_m^2(T_2-T_3)-(T_1-T_{amb})}+\frac{N^{TE}}{N^{GT}}\frac{\eta_m}{\eta_m^2(T_6-T_7)-(T_5-T_{amb})}\right](T_8-395)
\tag{4.27}
$$

$$
B = \left(1+\frac{N^{TE}}{N^{GT}}\right)
\tag{4.28}
$$

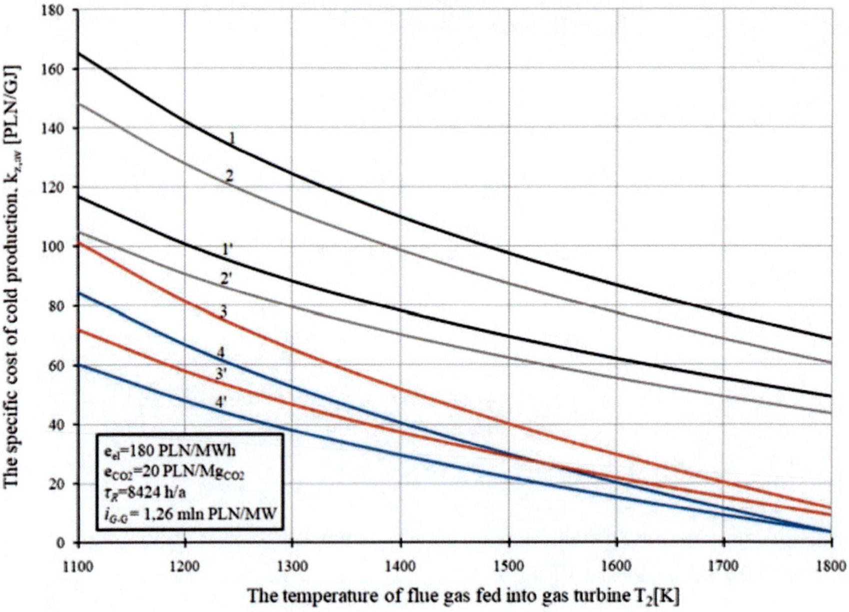

Fig. 4.12 The values of unit cooling production costs $k_{z,av}$ in the gas-gas system with the turboexpander with no heat recovery with the thermal absorption chiller as the function of the temperature of inlet exhaust gases to the gas turbine T_2

$$i_z = \frac{J_z}{\dot{Q}_z} \tag{4.29}$$

Figures 4.12, 4.13, 4.14, 4.15, 4.16 and 4.17 present the results of thermodynamic calculations for the unit cooling production cost $k_{z,av}$ for the *gas-gas* system with the turboexpander without heat recovery. Table 4.3 lists calculation input data for individual curves presented in Figs. 4.12, 4.13, 4.14, 4.15, 4.16 and 4.17.

The calculation results presented in Figs. 4.12, 4.13, 4.14, 4.15, 4.16 and 4.17 show that the unit cooling production cost in the thermal absorption chiller depends mainly on the electricity price. Here, it should be noted that when we compare the cost with the cost of cooling production in the system with the compressor chiller for value $x = 0.1$, i.e. the same for both cooling production systems, then the unit cost is lower for the compressor chiller.

Calculating the cost $k_{z,av}$ in the system with heat recovery in the turboexpander, Fig. 4.2, assuming that investment outlays on the system with regeneration are the same as for the system without regeneration, it turns out that reducing the cost, similarly to the system with the compressor chiller, is negligible and does not exceed the value of 1 PLN/GJ within the whole temperature range T_2.

To increase the cooling power of the absorption chiller (4.21), we should increase the temperature range of exhaust gases $\Delta T = T_8 - 395$ that drives it. The range,

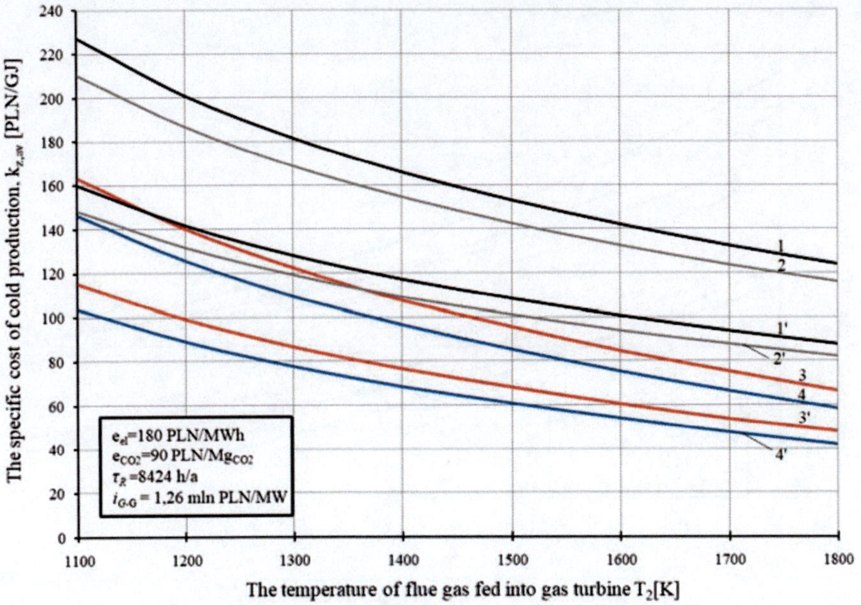

Fig. 4.13 The values of unit cooling production costs $k_{z,av}$ in the gas-gas system with the turboexpander with no heat recovery with the thermal absorption chiller as the function of the temperature of inlet exhaust gases to the gas turbine T_2

as already emphasised, is relatively low and depending on the temperature of inlet exhaust gases delivered to the gas turbine it ranges T_2 from approx. 63 K to approx. 130 K. To increase this range, it would be necessary to additionally burn gas (so-called supplementary firing) in the turboexpander system in the space of a duct supplying air from the heater N to the turboexpander TE (this is known from *gas-steam* systems process of the so-called afterburn [5, 6])—Fig. 4.1c. Then the temperature T_6 will be increased, and a as a result, the temperature T_8. However, it should be emphasised that burning fuel using burners placed in the afterburn chamber located in the N-TE channel of the turboexpander causes that the hierarchical engine—Figs. 4.1a, b, c and 4.2—becomes the quasi-hierarchical engine, i.e. a dual engine [6–8]. The energy efficiency values, especially for exergy quasi-hierarchical systems, are lower than the efficiency of hierarchical systems.

Increasing temperature T_6 with the value $\Delta T_6 = 100$ K by burning gas in the N-TE channel, T_8 increases to the value of T_8^{sup}. Depending on the value of temperature $T_2 \in \langle 1100;\ 1800\,\text{K}\rangle$ temperature T_8 from value $T_8 \in \langle 458;\ 524\,\text{K}\rangle$ increases to value $T_8^{\text{sup}} \in \langle 487;\ 555\,\text{K}\rangle$. Increasing T_6 by $\Delta T_6 = 200$ K the temperature is $T_8^{\text{sup}} \in \langle 517;\ 586\,\text{K}\rangle$. The ratio of cooling power to the system electric power grows then from the value with no gas afterburn feature $\dot{Q}_z / N_{el}^{G-G} = \langle 0,32;\ 0,23\rangle$, Fig. 4.11, to the value of $\dot{Q}_z^{\text{sup}} / N_{el}^{G-G} = \langle 0,38;\ 0,25\rangle$ at the temperature increase T_6 by value

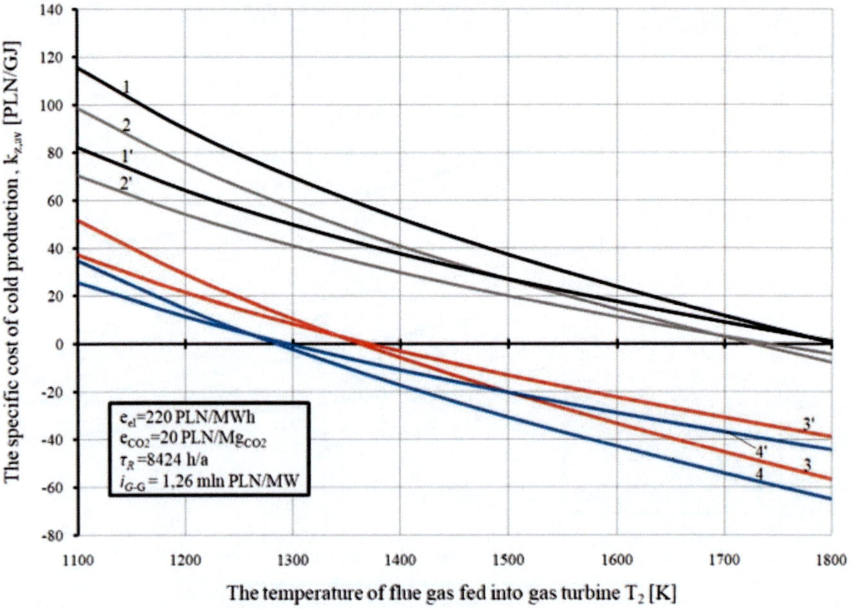

Fig. 4.14 The values of unit cooling production costs $k_{z,av}$ in the gas-gas system with the turboexpander with no heat recovery with the thermal absorption chiller as the function of the temperature of inlet exhaust gases to the gas turbine T_2

$\Delta T_6 = 100$ K and to value $\dot{Q}_z^{\text{sup}}/N_{el}^{G-G} = \langle 0,43;\ 0,26\rangle$ at the temperature increase T_6 by value $\Delta T_6 = 200$ K.

As a result of the temperature increase T_6 by value ΔT_6 grows the efficiency $\eta_{TE}^{\max}$ to value $\eta_{TE}^{\max,\,\text{sup}}$. The value of those efficiencies can be calculated from formula (4.43) after substituting there for η_R zero and for temperature T_{6R} values T_6 and T_6^{sup}, respectively. Values $\eta_{TE}^{\max}$ ($\eta_{TE}^{\max}$ is calculated obviously for $\Delta T_6 = 0$) and $\eta_{TE}^{\max,\text{sup}}$ are presented in Fig. 4.19.

Burning additional fuel in the turboexpander changes the efficiency of electricity generation for this turboexpander. It can be determined from the energy balance:

wherein:

$\dot{I}_8^{\text{sup}}$ enthalpy of flue gas fed into the chiller following supplementary firing of the fuel,

q level of supplementary firing in the engine, $q = \dot{Q}_{\text{sup}}/\dot{E}_{ch}$,

from which the formula for the efficiency of electricity generation in the *gas-gas* engine with afterburn is determined:

$$\eta_{G-G}^{\text{sup}} = \frac{N^{GT} + N^{TE,\text{sup}}}{\dot{E}_{ch} + \dot{Q}_{\text{sup}}} = \frac{\eta_{GT}^{\max} + \eta_{TE}^{\max,\text{sup}} - \eta_{GT}^{\max}\eta_{TE}^{\max,\text{sup}} + q\eta_{TE}^{\max,\text{sup}}}{1+q} \tag{4.30}$$

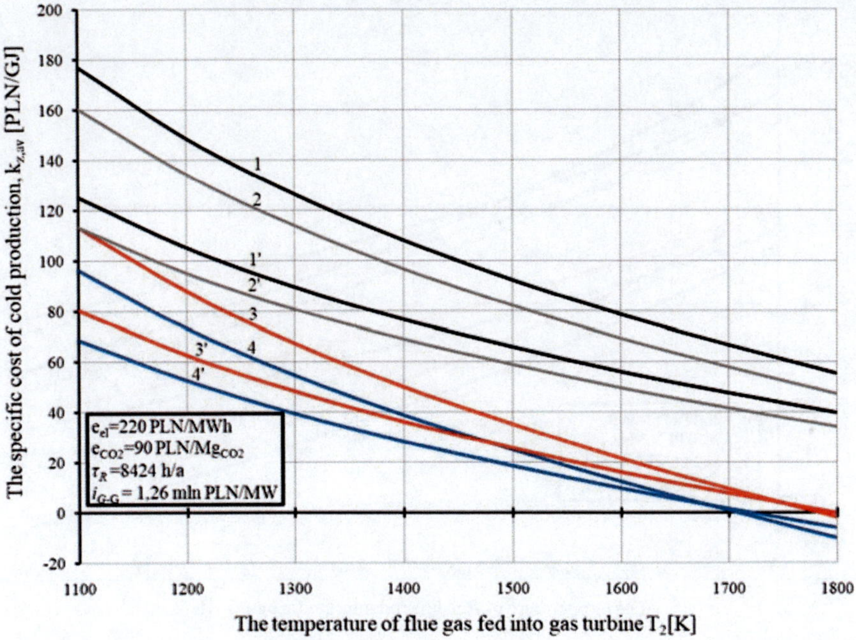

Fig. 4.15 The values of unit cooling production costs $k_{z,av}$ in the gas-gas system with the turboexpander with no heat recovery with the thermal absorption chiller as the function of the temperature of inlet exhaust gases to the gas turbine T_2

and if $q = 0$ is assumed, then equation (4.30) presents the efficiency of the *gas-gas* engine without afterburn feature (refer to formula (2.23)):

$$\eta_{G-G} = \eta_{GT}^{\max} + \eta_{TE}^{\max} - \eta_{GT}^{\max}\eta_{TE}^{\max} \tag{4.31}$$

The values of efficiency η_{G-G} and η_{G-G}^{sup} as the function of the temperature of exhaust gases delivered to the gas turbine T_2 and value ΔT_6 are presented in Fig. 4.19. Dividing the dependency (4.30) by (4.31) we obtain:

$$\frac{\eta_{G-G}^{\text{sup}}}{\eta_{G-G}} = \frac{1}{1+q}\,\frac{1 - \eta_{TE}^{\max,\text{sup}} + \frac{\eta_{TE}^{\max,\text{sup}}}{\eta_{GT}^{\max}}(1+q)}{1 + \frac{\eta_{TE}^{\max}}{\eta_{GT}^{\max}} - \eta_{TE}^{\max}} \tag{4.32}$$

The afterburn fuel chemical energy stream can be determined from formula:

$$\dot{Q}_{\text{sup}} = \dot{P}_{\text{sup}} W_d = \dot{C}_{TE}(T_6^{\text{sup}} - T_6) \tag{4.33}$$

where:

$\dot{P}_{\text{sup}}$ afterburn gas mass stream,

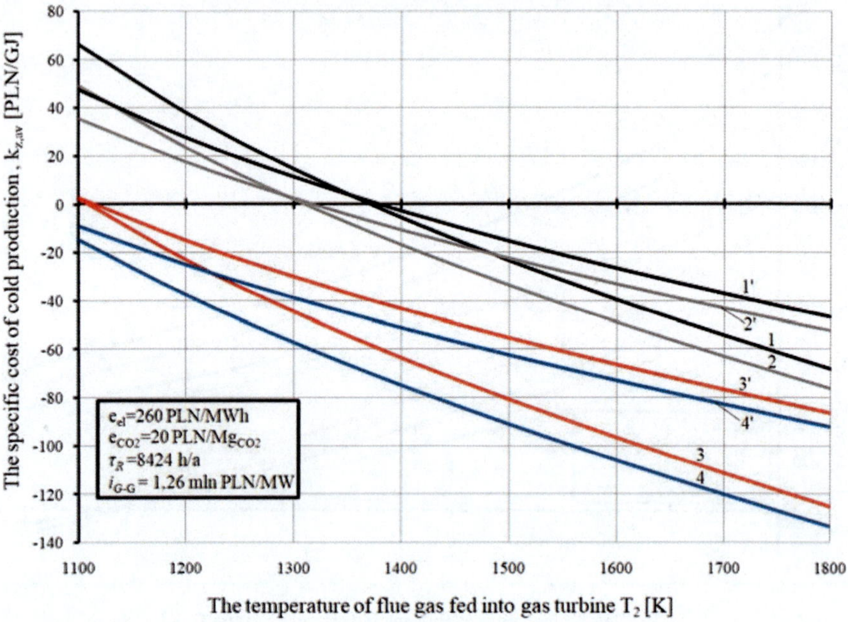

Fig. 4.16 The values of unit cooling production costs $k_{z,av}$ in the gas-gas system with the turboexpander with no heat recovery with the thermal absorption chiller as the function of the temperature of inlet exhaust gases to the gas turbine T_2

W_d gas calorific value,

wherein the thermal capacity stream of exhaust gases $\dot{C}_{TE}$ delivered to the turboexpander equals approximately the air thermal capacity before the afterburn process, which in turn, which results from the balance of the heater N, equals approximately the thermal capacity stream of exhaust gases $\dot{C}_{GT}$ removed from the gas turbine (see formula (3.15)). The capacity stream $\dot{C}_{GT}$ results from the turbine energy balance—Fig. 4.18:

$$\dot{E}_{ch} - N^{GT} = \dot{E}_{ch}(1 - \eta_{GT}^{\max}) = \dot{I}_3 = \dot{C}_{GT}(T_3 - T_{amb}) \tag{4.34}$$

where:

$\dot{I}_3$ enthalpy stream of exhaust gases from the gas turbine,

Using Eqs. (4.33) and (4.34) from dependency $\dot{C}_{TE} \approx \dot{C}_{GT}$,we can determine the level of fuel afterburn:

$$q = \frac{\dot{Q}_{\text{sup}}}{\dot{E}_{ch}} = (1 - \eta_{GT}^{\max})\frac{T_6^{\text{sup}} - T_6}{T_3 - T_{amb}} \tag{4.35}$$

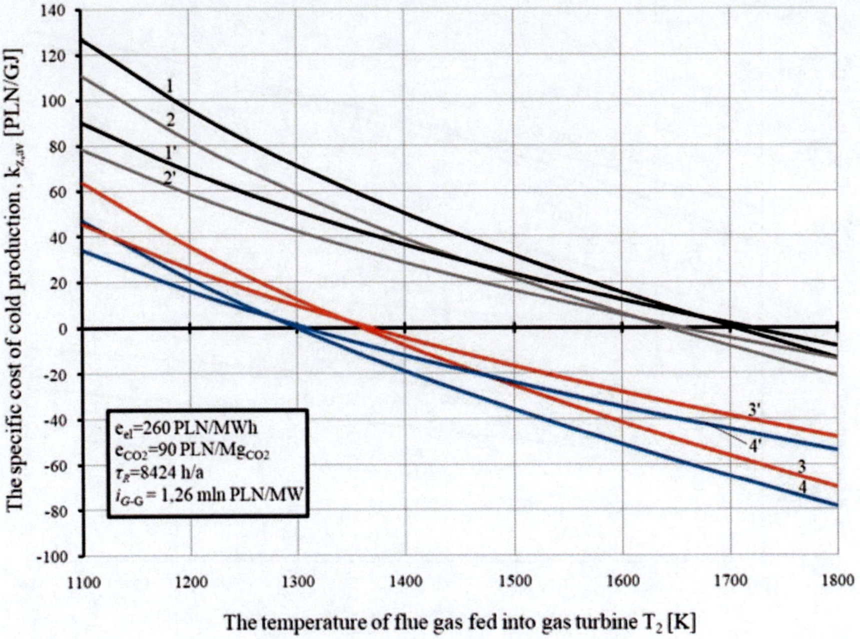

Fig. 4.17 The values of unit cooling production costs $k_{z,av}$ in the gas-gas system with the turboexpander with no heat recovery with the thermal absorption chiller as the function of the temperature of inlet exhaust gases to the gas turbine T_2

Table 4.3. Input data list

	1	1'	2	2'	3	3'	4	4'
e_c, PLN/GJ	50	50	70	70	50	50	70	70
e_{fuel}, PLN/GJ	24	24	24	24	20	20	20	20
i_z, PLN/MW	$0,5i_{G-G}$	$0,7i_{G-G}$	$0,5i_{G-G}$	$0,7i_{G-G}$	$0,5i_{G-G}$	$0,7i_{G-G}$	$0,5i_{G-G}$	$0,7i_{G-G}$
τ_z, h/a	$0,7\tau_A$	$\tau_z = \tau_A$	$0,7\tau_A$	$\tau_z = \tau_A$	$0,7\tau_A$	$\tau_z = \tau_A$	$0,7\tau_A$	$\tau_z = \tau_A$

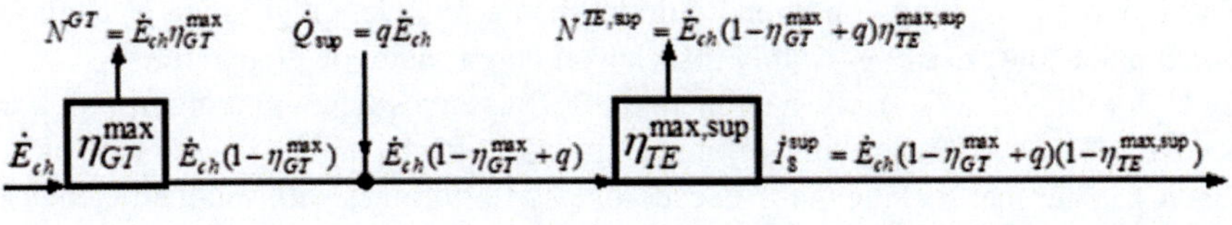

Fig. 4.18 The energy balance of the hierarchical gas-gas engine with supplementary firing

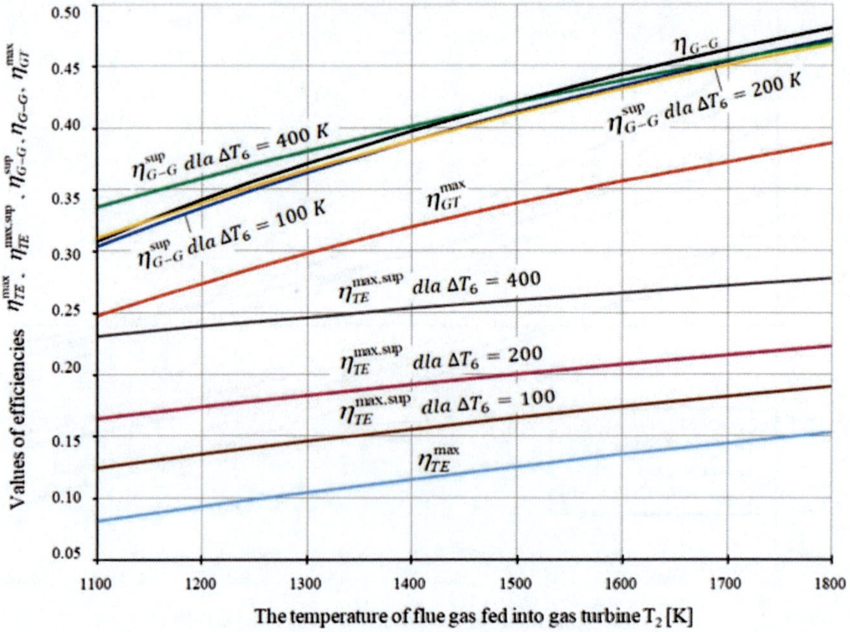

Fig. 4.19 The values of efficiency η_{TE}^{max}, $\eta_{TE}^{max,sup}$, η_{GT}^{max}, η_{G-G}, η_{G-G}^{sup} as the function of the temperature of exhaust gases delivered to the gas turbine T_2 for temperature increase $\Delta T_6 = T_6^{sup} - T_6$ of $\Delta T_6 = 0$, 100, 200, 400 K

The afterburn, in spite of increasing the value η_{TE}^{max} to the value $\eta_{TE}^{max,sup}$—Fig. 4.18.—reduces the efficiency of electricity generation in the engine. The value of relation is $\eta_{G-G}^{sup}/\eta_{G-G}$, though slightly, but lower than one, because the efficiency η_{G-G}^{sup} is slightly lower than η_{G-G}—Fig. 4.19. Here, it should be stressed that the revenue from selling electricity has a decisive impact on economic viability of the trigeneration system. Therefore, burning additional gas in the N-TE channel is not only, thermodynamically unviable, but also because of the low increase in chilling power of the system and increase in investment outlays related to afterburn burners and the combustion chamber in the N-TE channel space is economically unviable. The economic viability condition could be then a relatively high price of cooling power and a low price of gas. Only for relatively high values of temperature T_6^{sup}, i.e. for high values of ΔT_6, so when temperature T_6^{sup} assumes values ranging from 1,100 K to 1,400 K equalling values of temperature T_2 of inlet exhaust gases for “old”, that is gas turbines that are many-decade-old, so for turbines with small efficiencies η_{GT}, relation $\eta_{G-G}^{sup}/\eta_{G-G}$ is, though slightly, but higher than one. Low efficiency values for old turbine designs result from low heat-resistance of turbine blades, which forces low allowable temperature values T_2. Current designs allow temperatures T_2 exceeding even 1,800 K. However, a high afterburn level in the turboexpander causes that the hierarchical engine with the turboexpander stops making technical sense. The turboexpander then becomes in some way the gas turbine.

The same curve characteristics, as curves in Fig. 4.19, have curves $\eta_{TE,R}^{max}$, $\eta_{TE,R}^{max,sup}$ for the engine with heat recovery in the turboexpander—Fig. 4.2—when the gas afterburn feature is used in the channel N-TE.

4.3 Using a Turboexpander with Heat Recovery in a Trigeneration Gas-Gas System

Calculating the energy balance of *Joule-Brayton* cycle for the turboexpander operating with the regeneration heat exchanger *R*—Figs. 4.2, 4.20—we obtain formula for its energy efficiency [1]:

$$\eta_{TE} = \frac{N_{TE,R}}{\dot{Q}_{in}} = \frac{N_{i\ \exp} - N_{i\ C}}{\dot{Q}_{in}} = \frac{\eta_m (T_{6R} - T_{7R}) - \frac{1}{\eta_m}(T_{5R} - T_{amb})}{T_{6R} - T_{NR}} \tag{4.36}$$

After substituting to formula (4.36), the temperatures of heat transfer medium

$$T_{5R} = T_{amb} + \frac{1}{\eta_i^C}(T_{5Rs} - T_{amb}) \tag{4.37}$$

$$T_{7R} = T_{6R} - \eta_i^{TE}(T_{6R} - T_{7Rs}) \tag{4.38}$$

$$T_{NR} = (1 - \eta_R)T_{5R} + \eta_R T_{7R} \tag{4.39}$$

$$T_{5Rs} = T_{ot} z_{TE,R} \tag{4.40}$$

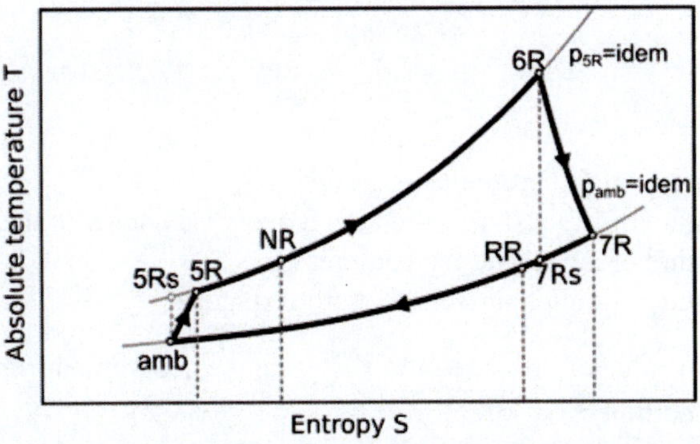

Fig. 4.20 The *Joule-Brayton* cycle of the turboexpander with heat recovery (recovery heat equals the surface area under the RR-7R isobar section and the entropy axis); the surface area equals the surface area under the 5R-NR isobar section (the following relations are used: $T_{7R} > T_{NR}$, $T_{RR} > T_{5R}$)

$$T_{7Rs} = T_{6R} / z_{TE,R} \tag{4.41}$$

where:

$$z_{TE,R} = \left(\frac{p_{5R}}{p_{amb}}\right)^{\frac{\kappa-1}{\kappa}} = \left(\frac{p_{6R}}{p_{7R}}\right)^{\frac{\kappa-1}{\kappa}} \tag{4.42}$$

we obtain its final form (for values $z_{TE,R}^{opt}$ determined from formula (4.46)):

$$\eta_{TE,R}^{\max} = \frac{N_{TE,R}^{\max}}{\dot{Q}_{in}} = \frac{\eta_m \eta_i^{TE} T_{6R}\left(1 - \frac{1}{z_{TE,R}^{opt}}\right) - \frac{1}{\eta_m \eta_i^C} T_{amb}(z_{TE,R}^{opt} - 1)}{T_{6R} - (1 - \eta_R) T_{amb}\left[1 + \frac{1}{\eta_i^C}(z_{TE,R}^{opt} - 1)\right] - \eta_R T_{6R}\left[1 - \eta_i^{TE}\left(1 - \frac{1}{z_{TE,R}^{opt}}\right)\right]} \tag{4.43}$$

Substituting in formula (4.43) for η_R the value of zero, $\eta_R = 0$, we obtain the formula for energy efficiency of *Joule-Brayton* cycle for the turboexpander without recovery (compare to formula (4.1)). In turn, when to (4.31) for $\eta_{TE}^{\max}$ we substitute obtained from (4.43) value $\eta_{TE,R}^{\max}$, then we obtain a formula for efficiency $\eta_{G-G,R}$ (Fig. 4.23) of the *gas-gas* engine with the turboexpander operating with heat recovery.

The temperature of inlet air delivered to the turboexpander T_{6R} can be determined for a known value T_3 (4.11) and the assumed value ΔT_1:

$$T_{6R} = T_3 - \Delta T_1 \tag{4.44}$$

Effectiveness η_R of the heat exchanger *R*, Fig. 4.2, can be defined by the dependency:

$$\eta_R = \frac{T_{NR} - T_{5R}}{T_{7R} - T_{5R}} = \frac{T_{7R} - T_{RR}}{T_{7R} - T_{5R}} \tag{4.45}$$

where:

T_{NR} temperature of air before the air heater,
T_{RR} temperature of exhaust air from the recovery exchanger to the stack,
T_{5R} temperature of air behind the compressor,
T_{7R} temperature of outlet air from the turboexpander.

The optimum value of pressure ratio $z_{TE,R}^{opt}$ can be determined from the equation below (from condition $d\eta_{TE} / dz_{TE,R} = 0$):

$$[(a+b)c - ad](z_{TE,R}^{opt})^2 + 2(ae - bc)\, z_{TE,R}^{opt} - (a+b)e + db = 0 \tag{4.46}$$

where:

$$a = \frac{T_{amb}}{\eta_m \eta_i^C}, b = \eta_m \eta_i^{TE} T_{6R}, c = \frac{1 - \eta_R}{\eta_i^C} T_{amb},$$

$$d = T_{6R} + (1 - \eta_R) T_{amb} \left(\frac{1}{\eta_i^S} - 1 \right) - \eta_R T_{6R} (1 - \eta_i^{TE})$$

$$e = \eta_R \eta_i^{TE} T_{6R} \tag{4.47}$$

The negative root of Eq. (4.46) does not have a physical meaning.
Knowing value $z_{TE,R}^{opt}$ from formula:

$$z_{TE,R}^{opt} = \frac{T_{5Rs}}{T_{ot}} = \frac{T_{6R}}{T_{7Rs}} = \left(\frac{p_{5R}}{p_{amb}} \right)_{opt}^{\frac{\kappa-1}{\kappa}} \tag{4.48}$$

we can determine the value p_{5R}^{opt} and by using irreversible adiabatic process, we can calculate the temperature of air behind the compressor T_{5R} and the turboexpander T_{7R}:

$$T_{5R} = T_{amb} \left[1 + \frac{1}{\eta_i^C} (z_{TE,R}^{opt} - 1) \right] \tag{4.49}$$

$$T_{7R} = T_{6R} \left[1 - \eta_i^{TG} \left(1 - \frac{1}{z_{TE,R}^{opt}} \right) \right] \tag{4.50}$$

Knowing temperatures T_{5R} and T_{7R} using equation (4.45), we can determine temperatures T_{NR}, T_{RR}.

Figures 4.21, 4.22 and 4.23 present the results of thermodynamic calculations for the *gas-gas* system operating with the turboexpander with the recovery heat exchanger—Figs. 4.2, 4.20—and for comparative purposes for the system without regeneration.

Figures 4.21, 4.22 and 4.23 present the calculation results showing that using heat recovery in the turboexpander, Fig. 4.2, is thermodynamically justified. This is because the efficiencies $\eta_{TE,R}^{\max}$ increase and thus $\eta_{G-G,R}$, so the power of the turboexpander increases, including electric power of the whole system. The ratio of turboexpander power to gas turbine power is then approx. $N^{TE}/N^{GT} \approx 15.5\%$ when in the system without recovery—Fig. 4.1a, b, c—this ratio is approx. $N^{TE}/N^{GT} \approx 15.5\%$. The temperature of exhaust gases, though only by approx. 1 °C, also increases T_{8R} over T_8 (temperatures "merge" in Fig. 4.21), which almost does not increase the cooling power of the absorption chiller (formula (4.21)).

It is important to answer the question whether the *gas-gas* system with the turboexpander and recovery heat exchanger is not only thermodynamically viable, but also economically viable The prerequisite for this is that level of revenue obtained from selling additional electricity generated in the turboexpander, as well as cooling in the

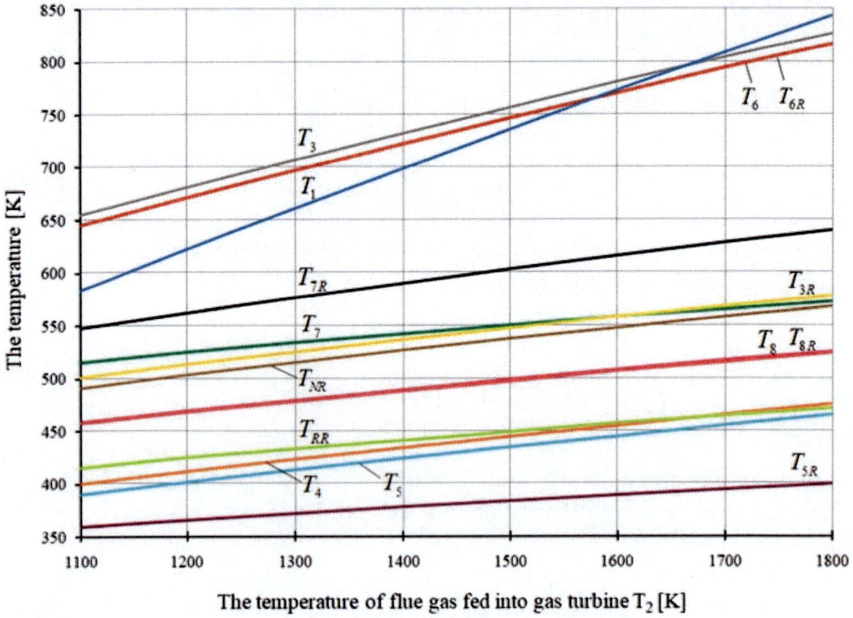

Fig. 4.21 Temperature values T_1,T_3,T_4,T_5,T_6,T_7,T_8 and T_{NR},T_{RR}, T_{3R},T_{5R},T_{6R},T_{7R},T_{8R} as the function of the temperature of inlet exhaust gases delivered to the gas turbine T_2

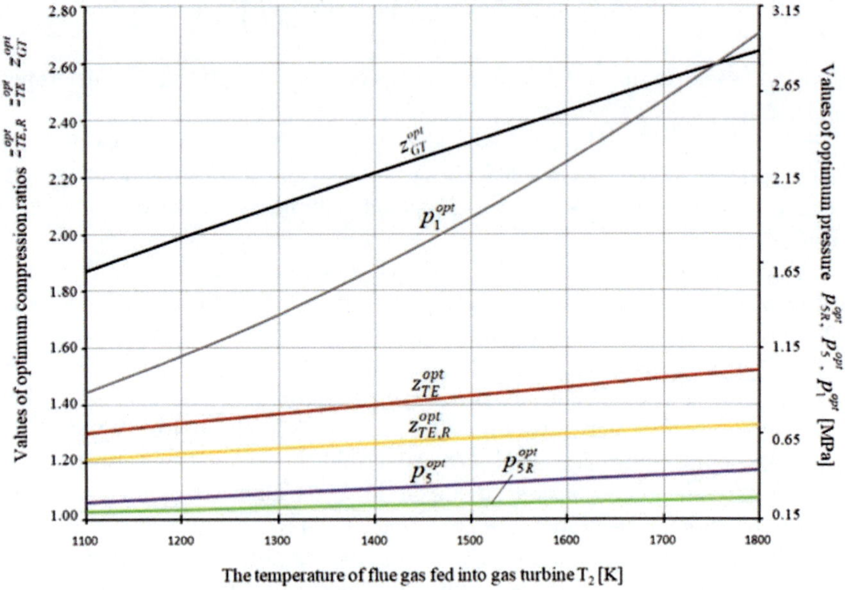

Fig. 4.22 Values of optimum pressures p_1^{opt},p_5^{opt},p_{5R}^{opt} and optimum compression ratios z_{TE}^{opt},$z_{TE,R}^{opt}$, z_{TG}^{opt} as the function of temperature of exhaust gases delivered to the gas turbine T_2

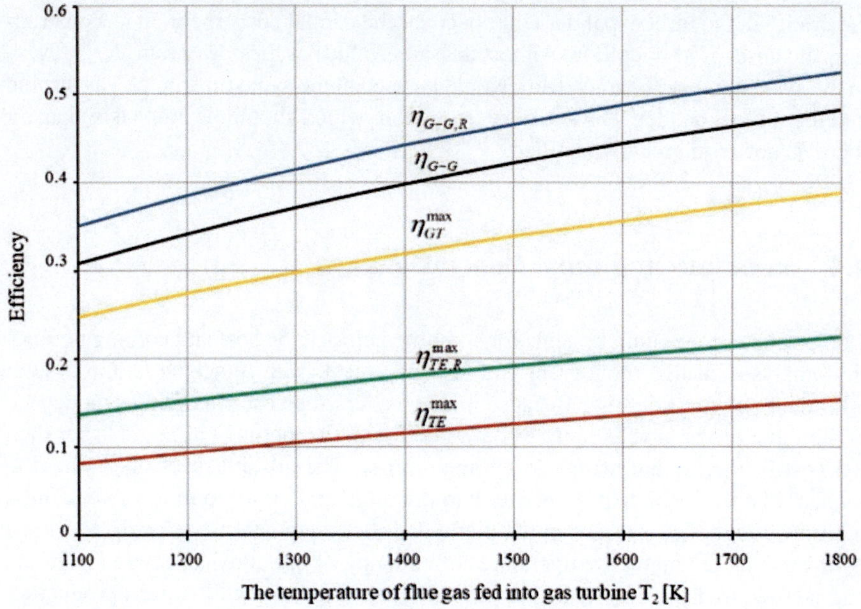

Fig. 4.23 Efficiency values η_{G-G},$\eta_{G-G,R}$, η_{TE}^{max}, $\eta_{TE,R}^{max}$, η_{TG}^{max} as the function of the temperature of exhaust gases delivered to the gas turbine T_2

gas-gas system, are not lower than the increase in the annual capital costs (depreciation and financial costs) and the operating costs (maintenance and repair costs) [2–4] related to higher investment outlays on the heat regenerator. Assuming event, as already mentioned, that the increase in investment outlays is null, i.e. when the outlays are the same as on the system without the regenerator, reducing the unit cooling production cost is negligible, does not exceed the value of 1 PLN/GJ in the case of the thermal chiller and 0.5 PLN/GJ in the case of the compressor chiller. If the calculations took into account the increase, then the unit cost at the best-case scenario would be zero. It means that using regeneration is not economically viable. Then, it can be concluded that the design of the *gas-gas* system with the turboexpander operating with heat recovery, both in the system with the compressor and thermal chiller, is completely economically unviable.

Also the use of heat regeneration in the gas turbine is not economically justified. Then, the temperature of exhaust gases from the gas turbine regeneration exchanger is relatively low, lower by approx. 100 K from the temperature T_3 of exhaust gases removed from the turbine without recovery—Figs. 4.1, 4.2. As a result, the air heated in the heater N with these exhaust gases driving the turboexpander cause its relatively low power that ranges then from barely 5.5% to 8% of the gas turboexpander (values 5.5% and 8% respectively correspond to the temperatures T_2 of inlet exhaust gases delivered to the gas turbine of 1,100 and 1,800 K). It means that the production of electricity in the turboexpander will not cover the capital costs spent on its construction and costs of its operation. Almost the same value has the power ratio N^{TE}/N^{GT} in the system, where heat recovery would be used at the same time in the gas turbine and the turboexpander. The use of regeneration, which should be emphasised again there, is not economically justified.

4.4 Summary and Final Conclusions

Comparing trigeneration systems for combined electricity, heat and cooling production analysed in this monograph, more economically advantageous is the system with the compressor chiller Because in this system the cost of cooling production, at the same production as in the system fitted with the thermal chiller, is lower than the cost in the system using the thermal chiller. The advantage of this system is the fact that thanks to using electricity to drive the chiller, its cooling power can be arbitrarily high. By contrast, in the thermal chiller the power is seriously restricted by the available range of temperatures of enthalpy of outlet exhaust gases from the engine used to driving it. In the system with the compressor chiller, heat production is also high, over 3 times higher than in the system with the thermal chiller and the revenue from selling it is the avoided cost of cooling production. It results from using the whole available temperature range of enthalpy of outlet exhaust gases from the engine, while in the system with the thermal chiller it is used at the same temperature range—Fig. 4.1a, c—also to provide cooling.

It should also be emphasised that electricity production in systems both with the compressor chiller and thermal chiller is higher, the higher is the efficiency of its generation in the gas turbine set and turboexpander, i.e. the higher the temperature of T_2 inlet exhaust gases to the turbine set. While the revenue from selling electricity, the same as the revenue from selling heat, is the avoided cost of cooling production. The increase in this temperature significantly reduces its unit production costs—Figs. 4.4, 4.5, 4.6, 4.7, 4.8, 4.9, 4.12, 4.13, 4.14, 4.15, 4.16 and 4.17. However, it should be emphasised that the revenue from electricity sale in the system with the compressor chiller reduces along with the increase in its cooling power. Therefore, the compressor chiller is, as already time and again emphasised, driven with electricity generated in the system.

In addition, it is important to answer the question whether using regeneration heat exchangers in the hierarchical *gas-gas* system, both in the turboexpander and gas turbine system, increasing the efficiency of generating electricity in the system is economically viable. The answer is unequivocal – no. This is because heat exchangers increase investment outlays on the engine, thus raising capital costs (depreciation and financial costs) and costs of engine operation (maintenance and repairs), while the increase in electricity production is small enough that reducing the unit cost of cooling production, both in the system with the thermal and compressor chiller would be, at the best-scenario, zero.

Finally, we should say *expressis verbis* that a big advantage, which should not be overestimated, of hierarchical *gas-gas* engines, both for producing solely electricity and for cogeneration and trigeneration purposes, is the capacity to build them in zones without water.

References

1. Szargut J (1998) Technical thermodynamics (in Polish: Termodynamika Techniczna). Wydawnictwo Politechniki Śląskiej, Gliwice
2. Bartnik R, Bartnik B (2014) Economic calculation in the energy sector (in Polish: Rachunek ekonomiczny w energetyce). WNT, Warszawa
3. Bartnik R, Buryn Z, Hnydiuk-Stefan A (2017) Investment strategy in heating and CHP. Mathematical Models, Springer, London
4. Bartnik R, Bartnik B, Hnydiuk-Stefan A (2016) Optimum investment strategy in the power industry. Mathematical Models, Springer, New York
5. Bartnik R (2009) Gas-steam power plants and combined heat and power plants. Energy and economic efficiency (in Polish: Elektrownie i elektrociepłownie gazowoparowe. Efektywność energetyczna i ekonomiczna), WNT, Warszawa (reprint 2012, 2017).
6. Bartnik R, Buryn Z, Hnydiuk-Stefan A, Skomudek W (2019) Dual-Fuel Gas-Steam Combined Heat and Power Plants (in Polish: Dwupaliwowe elektrownie i elektrociepłownie gazowoparowe. Efektywność energetyczna i ekonomiczna), WNT, Warszawa
7. Bartnik R, Buryn Z, Hnydiuk-Stefan A, Skomudek W (2020) Power plant retrofit and modernization (in Polish: Modernizacja elektrowni. Efektywność energetyczna i ekonomiczna), WNT, Warszawa
8. Bartnik R (2013) The modernization potential of gas turbines in the coal-fired power industry. Thermal and Economic Effectiveness, Springer, London
9. Feidt M (2018) Finite physical dimensions optimal thermodynamics 2—complex systems, ISTE Press, Elsevier
10. Sieniutycz S, Jezowski J (2009) Energy optimization in process systems. Elsevier Science

Chapter 5
Economic Analysis of Hydrogen Production in the Process of Water Electrolysis in a Gas–Gas Engine System

5.1 Introduction

The chapter includes an economic analysis of unit hydrogen production costs in the water electrolysis process in the system with the *gas–gas* engine—Fig. 5.1. But not only. For comparative purposes, we present also the values of the cost for all remaining power generation technologies. This is because electricity is the "fuel" used to produce hydrogen in the water electrolysis process, therefore it is important to find out the cost for all available technologies in order to know how to produce it as cheap as possible. We also present, importantly, results of the so-called renewable energy sources (*RES*). *RES* are currently playing an important role in the power industry, despite their relatively low electricity production capacity, while unit investment outlays are high—Table 5.1.

The problem of hydrogen production is important for two reasons. The first one is that the "defence" of conventional power units and electricity generated inside them against the already mentioned *RES*. It is because *RES* have a priority access to the National Grid System (NGS), so when the wind starts blowing or the sun begins shining the conventional power units have to be shut down or seriously reduce their power. It takes place hundreds of times a year. As a result, the power units, especially boilers, are subject to extensive wear and their operational time gets shorter. To avoid such operation modes and to provide stable work and even load, it is possible to use them to produce hydrogen during *RES* operation. Another reason is that it is necessary to increase hydrogen production to supply fuel cells used, which is extremely important, to convert fuel chemical energy directly into electric power. The theoretical efficiency of electricity generation is then 100%, not like in mechanical engines is restricted with the *Carnot* "corset", formula (2.5). It should be emphasized that this is a generally accepted approach. However, some researchers use a thermodynamic approach by analogy with the *Carnot* cycle [6]. This, approach is based on a more general formulation where the lower and upper source of heat is replaced by the lower and upper source of energy, in this case chemical energy [7]. Nevertheless, the real efficiency of cells is at least a dozen percent points higher than

R. Bartnik and T. W. Kowalczyk, *Hierarchical Gas-Gas Systems*,
Power Systems, https://doi.org/10.1007/978-3-030-69205-6_5

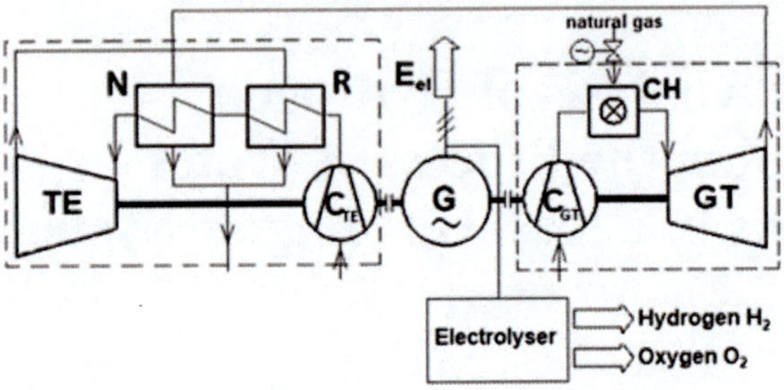

Fig. 5.1 System for hydrogen production in the electrolyser (dotted line rectangles mean that it is favourable when the turboexpander TE and low-pressure compressor C_{TE} are integrated in a single casing and share a common shaft as is the case for the turbine part and the gas turbine GT high-pressure compressor C_{GT})

efficiency of all remaining available power generation technologies, where at first in the chain of thermodynamic processes the fuel chemical energy is converted to heat, then changed to mechanical work and then to electric power, the most noble energy form, in the electric generator according to the *Faraday's* law.

Other possibilities of defending a conventional power plant against *RES* include storing electricity by using compressed air (a relevant problem can be caused by leaks of underground mine excavations used for storage purposes), magnetic fields, liquid air, acid-lead, nickel–cadmium and lithium-ion batteries. Methods are not justified in terms of technology and economy, while their "storage" capacities are extremely low. The most reasonable method, used for a long time, is storing electricity by pumping water in upper reservoirs of pumped-storage hydroelectricity plants.

Without defending power plants using fossil fuels against *RES* there will not be cheap electricity available in wall outlets constantly throughout the year, the electric power from *RES* will only be available on average for one hundred and a couple dozen hours a month at a horrendous price (Table 5.2). With a complete unpredictable time availability of the power. Is it possible for the world to "operate" like this? This question is, obviously, purely rhetorical.

As already mentioned, one of the methods for defending a traditional power plant against *RES* can be producing hydrogen in electrolysers in them. Therefore, it is necessary to analyse its cost. But not only. It is also necessary to analyse the cost of hydrogen production in *RES*. Because it is important to find the answer to the following question:

- which of the hydrogen production costs is lower: the cost of electric power from the power plant or from *RES*?

To answer the question, it is necessary to know unit electricity production costs in power plants fired by fossil fuels and in *RES*—Table 5.2 [4]. These costs, as already mentioned, translate into the hydrogen price, because it is electricity that is used as "fuel" to electrolyse water. The values of those unit costs for data from Table 5.1 are presented in Table 5.2.

Table 5.1 List of basic input data for calculations of unit costs for electricity generation in power generation technologies under analysis [4]

Power plant type	Coal-fired with air combustion	Coal-fired with oxy- combust ion	Nuclear	Photovoltaic	Prosumer photovoltaic	Wind	Prosumer wind	Combined cycle (CCPP)	Dual-fuel combined cycle (DFCC)
Estimated investment i, mln PLN/MW	6,5	9.1	18	6.3 (1.5 euro/W)	1 2. 6 (3 euro/W)	6.3 (1.5 euro/W)	12.6 (3 euro/W)	2.7	4.6
Annual operating time t_R, h/year	7500	7500	8000	750	750	1750	1750	7500	7500
Internal electrical load: ε_{el}, %	7.6	33	7.6	1	1	1	1	4	6.2
Construction period b, years	5	5	5	1	1	1	1	2	5
Specific fuel price, PLN/GJ	11.4	11.4	6.6	0	0	0	0	32	coal = 11.4 gas = 32
Exploitation period: $T = 20$ years									
Annual rate of maintenance and overhaul $d_{serv} = 3\%$									
Coefficients: $x_{sa}i x_{Ins} = 0.25$; $x_{swm,was} = 0.02$									
Discount rate $r = 8\%$									
CO_2 emission charges: $e_{CO2} = 29.4$ PLN/Mg_{CO2}, ($e_{CO2} = 7$ euro; exchange rate EURO/PLN = 4.2)									

(continued)

Table 5.1 (continued)

Power plant type	Coal-fired with air combustion	Coal-fired with oxy- combust ion	Nuclear	Photovoltaic	Prosumer photovoltaic	Wind	Prosumer wind	Combined cycle (CCPP)	Dual-fuel combined cycle (DFCC)
Tariff charges on emissions: p_{CO2} = 0,29 PLN/Mg CO2, P_{CO} = 110 PLN/MgCO, p_{NOX} = 530 PLN/MgNOx, P_{SO2} = 530 PLN/Mg SO2, p dust = 350 PLN/Mgdust									
Emission from coal combustion: p_{CO2} = 95 kg/GJ, p_{CO} = 0,01 kg/GJ, p_{NOx} = 0.164 kg/GJ, p_{SO2} = 0.056 kg/GJ, p_{dust} = 0.007 kg/GJ									
Emission on gas combustion: p_{CO2} = 55 kg/GJ, p_{CO} = 0 kg/GJ, p_{NOx} = 0.02 kg/GJ, p_{SO2} = 0 kg/GJ, p_{dust} = 0 kg/GJ									
Ratio of chemical energy of the fuel in its total annual use for which the purchase of additional allowances is not required: CO_2: u = 0									

(unit nuclear fuel price 6.6 PLN/GJ includes the cost of its disposal; makes approx. 20% of this price).

Table 5.2 List of average unit costs of electricity generation in power generation technologies under analysis [4] corresponding to base data from Table 5.1

Power plant type	Nuclear	Photovoltaic	Prosumer photovoltaic	Wind	Prosumer wind	Dual-fuel combined cycle (DFCC)	Combined cycle (CCPP)	Coal-fired with air combustion	Coal-fired with oxy-combustion $X_{ccs} = 0.2$
Specific cost of electric power generation $\hat{k}$ [PLN/MWh]	419	1217	2434	522	1043	296	276	279	463
Specific cost of electric power generation after depreciation period k_{eUmrt} [PLN/MWh]	115	318	636	136	273	214	234	160	232

Unit cost of electricity generation in the hierarchical *gas–gas engine*—Fig. 5.1—is almost the same as the cost in the hierarchical *gas-steam* technology—Figs. 3.22 and 3.23. Therefore, for the calculations of unit hydrogen production cost in the hierarchical engine system its value of 276 PLN/MWh was assumed—Table 5.2.

Additional description is required for costs of electricity production in nuclear power units. It results from the important role the nuclear technology plays and will be still playing. Because this technology is a "clean" power generation method that does not emit carbon dioxide In the future this will be thermonuclear fusion provided that the work carried out to master this technology in practice is successful. The humanity will be then in possession of an inexhaustible "clean" power source. However, since the 1950s we have been hearing that it is quite close and still we have nothing. Then, will it ever happen? Is it possible for humans to interfere into the structure of atom? For nuclear reactions only a controllable number of nuclei is split and the process takes place outside the very atoms by limiting the number of "attacking" neutrons by means of control rods. If humans could interfere in a controllable way into the atomic structure, it would probably provide us with unlimited capacity. The capacity to create new reality, a new world both in the physical and spiritual dimension. What would be the consequences? Would *Midas touch*, converting everything into gold, become then a real curse?

Figure 5.2 presents the values of the specific electricity production cost in a nuclear power plant in operating time T for a wide range of changes in unit investment outlays i and an interest rate r of investment capital with remaining basic data values as in Table 5.1.

Currently, we can obtain a percent rate for the investment capital for building nuclear power plants at a level even from approx. $r = 1\%$ to approx. $r = 3\%$. Loans with such low interest rates are provided thanks to the brokerage of the Export Credit Agency (ECA). Such agencies operate in countries using nuclear technologies, e.g. in France and the USA. However, the basic condition for obtaining such funds is using a loan to buy products and services in native countries, where ECA agencies operate. Government guarantees for an investor significantly reduce interest rates. Figure 5.2 shows that the unit electricity production cost in a nuclear power unit for rate $r = 3\%$ (and obviously, lower ones) and for the operating time $T = 60$ years (the time is currently normal for nuclear power units) amounts even much less than 200 PLN/MWh, i.e. is significantly lower than costs obtained for remaining technologies. However, if we were supposed to pay off a loan with an interest rate of $r = 3\%$ in the period not $T = 60$, but $T = 20$ years, then the sales price of electricity would have to be higher than approx. 280 PLN/MWh for unit investment outlays $i = 18$ million PLN/MW and over approx. 240 PLN/MWh for unit investment outlays $i = 15$ million PLN/MW—Fig. 5.2 (for rate $r = 8\%$ and $i = 18$ million PLN/MW the price would be higher than 419 PLN/MW—Table 5.2). After the period of T years, that is when the power plant is already fully depreciated, the unit electricity production cost would be only 115 PLN/MWh—Table 5.2 When maintaining the price of electricity at the level.

from years of loan payment the investor would obtain high profits, which would result in paying high taxes. To avoid this, it would have to "employ" in advance

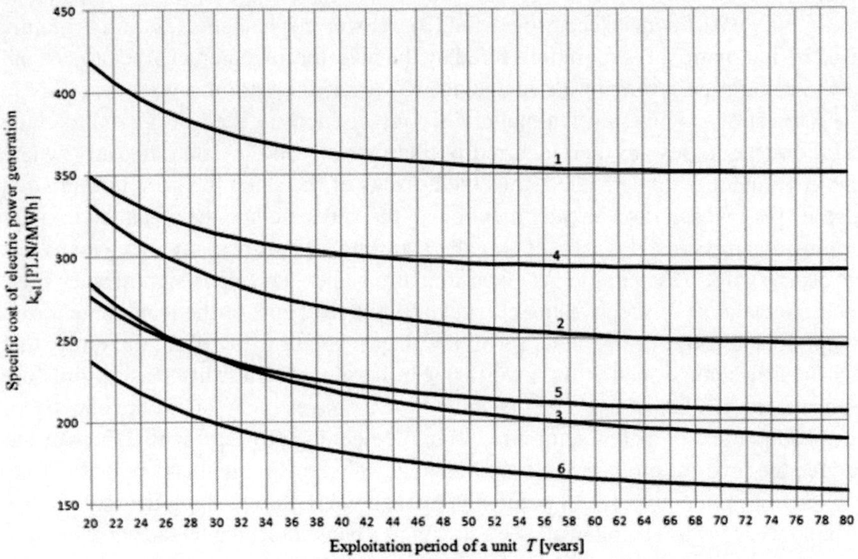

Fig. 5.2 Unit costs of electricity in a nuclear power station as the function of power unit operation time T for unit investment outlays i and a discount rate r for parameters: $1 - i = 18$ million PLN/MW, $r = 8\%$, $2 - i = 18$ million PLN/MW, $r = 5\%$, $3 - i = 18$ million PLN/MW, $r = 3\%$, $4 - i = 15$ million PLN/MW, $r = 8\%$, $5 - i = 15$ million PLN/MW, $r = 5\%$, $6 - i = 15$ million PLN/MW, $r = 3\%$,

methods of financial engineering to solve the problem and for all power unit construction and operation years develop a cash flow schedule i.e. revenues and all payments related to loan financial costs (i.e. the sum of its interests) and taxes, to maximise its profit. Here, it should be also noted that extending the chain of all payments, i.e. adding subsequent links to it always only results in losses. And this is not only a fundamental principle in economics, but also in all science branches; what is more, in all areas of life. By using a colloquial saying we can say that all sort of "combinations" bring always bad results, though it is also true that sometimes, according to the proverb, it turns out that *every cloud has a silver lining*. However, the worst thing in all this is definitely that in the interest of mighty people the world has always been ruled, is ruled and unfortunately certainly will be ruled by a lie, while the truth, not to say more emphatically, propaganda, is exclusively the one that makes the lie reliable.

Now we should go back to the high profit tax that has already been mentioned. As it should be expected, and even said with certainty, because it is obvious, the most favourable is to pay off the cheapest loan as fast as possible; additionally, collected, in subsequent instalments during the investment, as it minimises the financial cost of the loan, and then after paying it off, the sales price of electricity should be reduced. This should be the final result of analyses maximising the investor profit performed by using all the methods offered by the financial engineering that compliments financing any investment. The group includes three ways: (1) revolving, (2) deposits (plainly speaking, quasi-financial securities) and (3) transfer mechanisms. The ways mainly used by international corporations hired by the investor to implement the project and at the same time arranging its financing. However, they use the ways not only in the interest of the investor, but maliciously just to increase their profit and to elude taxation (the losses are then incurred by the investor and the state treasury where the investment is implemented, thus the citizens of the state: the state treasury can also be the investor). As already mentioned, the ways include the so-called transfer pricing mechanisms; it is also the so-called "rolling" subsequent revolving loans, i.e. revolving working capital loans taken in many banks. The loans significantly raise the financial cost of the investment (standard financial cost of the loan ranges from about 25% to 50% of its value, revolving may increase it to 100%; therefore, this is a fatal weapon aimed at the investor). Furthermore, banks that make profits are as a rule not the banks based in the country of the investment, but in the country of the headquarters of a contractor hired to implement the project. Finally, the ways to reduce the investor profit include various financial securities demanded by the party implementing the project. As a rule, the ordering party has to deposit some money, usually, very big, on an interest-free account in a bank indicated by the party hired to implement the project (obviously, the contractor collects interests from the deposited money). In addition, the mechanisms of transfer prices allow foreign corporations, international companies to transfer profits abroad. Wherever the state is inefficient and the law "leaky", they artificially manipulate profits and costs within a corporation so that profits and losses can be transferred where it is more convenient for them. They transfer illegal profits abroad e.g. in the form of payments being obviously fictitious costs to their headquarters in the form of licence fees, exorbitant fees for legal, technical, or financial expert opinions, export money from the country where they implement an investment cheaply, import expensively etc. As already mentioned, losses are always incurred by the state treasury and citizens. The electricity consumers are the ones who finally pay the electricity prices including all the unjustified production costs pushed up very high. If something has to be done well, or, more emphatically, honestly, spending funds in a justified way, it should be done according to the following principle: *Do it yourself!* If the investor cannot afford to implement an investment on their own, they have to face the fact that the investment will be expensive. Considering the financial engineering "tricks", even twice as expensive than it would be if it were implemented by the very investor, because the sum of financial costs of all loans will equal investment outlays. Hiring a foreign party to build a nuclear power station that at the same time arranges financing is the worst scenario for its construction since the most expensive one. It also obviously applies to any other investment.

5.2 Methodology and Mathematical Model for Calculating Unit Hydrogen Production Cost

When installing the electrolyser in the power unit taking in time t_{RES} electric power N_{el}^{RES} to produce hydrogen, then the profit of the power unit can be expressed by the following formula:

$$\begin{aligned}
NPV_{block}^{H_2} = \Bigg\{ & \Big[N_{el}t_A(1-\varepsilon_{el}) - N_{el}^{RES}t_{RES}\Big]\frac{e_{el}^{t=0}}{a_{el}-r}[e^{(a_{el}-r)T}-1] + m_{H_2}\frac{e_{H_2}^{t=0}}{a_{H_2}-r}[e^{(a_{H_2}-r)T}-1]+ \\
& -(1+x_{sw,m,was})\frac{N_{el}t_A}{\eta_{el}}\frac{e_{gas}^{t=0}}{a_{gas}-r}[e^{(a_{gas}-r)T}-1]+ \\
& -\frac{N_{el}t_A}{\eta_{el}}\frac{\rho_{CO_2}p_{CO_2}^{t=0}}{a_{CO_2}-r}[e^{(a_{CO_2}-r)T}-1] - \frac{N_{el}t_A}{\eta_{el}}\frac{\rho_{CO}p_{CO}^{t=0}}{a_{CO}-r}[e^{(a_{CO}-r)T}-1]+ \\
& -\frac{N_{el}t_A}{\eta_{el}}\frac{\rho_{NO_X}p_{NO_X}^{t=0}}{a_{NO_X}-r}[e^{(a_{NO_X}-r)T}-1] - \frac{N_{el}t_A}{\eta_{el}}\frac{\rho_{SO_2}p_{SO_2}^{t=0}}{a_{SO_2}-r}[e^{(a_{SO_2}-r)T}-1]+ \\
& -\frac{N_{el}t_A}{\eta_{el}}\frac{\rho_{dust}p_{dust}^{t=0}}{a_{dust}-r}[e^{(a_{dust}-r)T}-1] - \frac{N_{el}t_A}{\eta_{el}}(1-u)\frac{\rho_{CO_2}e_{CO_2}^{t=0}}{b_{CO_2}-r}[e^{(b_{CO_2}-r)T}-1]+ \\
& -(1+x_{sal,t,ins})(J+J_{SOE})(1-e^{-rT})\frac{\delta_{serv}}{r} - (zJ+J_{SOE})\left(\frac{1-e^{-rT}}{T}+1\right)\Bigg\}(1-p)
\end{aligned} \tag{5.1}$$

where:

$a_{el}, a_{gas}, a_{CO_2}, a_{CO}, a_{H_2}, a_{SO_2}, a_{NO_X}, a_{dust}, b_{CO_2}$	control [4] (exponentials),
$e_{el}, e_{gas}, e_{CO_2}\ e_{H_2}$	variable in time prices of unit electricity, fuel, CO_2 emission permission purchase and hydrogen; changes in time for those prices were assumed using exponentials, for example $e_{el}(t) = e_{el}^{t=0}e^{a_{el}t}$, $e_{H_2}(t) = e_{H_2}^{t=0}e^{aH_2t}\ e_{gas}(t) = e_{gas}^{t=0}e^{a_{gas}t}$,
δ_{serv}	annual fixed cost rate depending on investment outlays (costs of maintenance, repairs of devices).
ε_{el}	index of own power plant electric needs (its worth depends on electricity production technology used),
η_{el}	gross energy efficiency for electricity generation (its worth depends on technology used),
J, J_{SOE}	investment outlays on the power plant and elektrolyser, respectively,
N_{el}	gross electric power plant power,
u	share of fuel chemical energy in its total annual consumption, for which the purchase of required CO_2 emission permissions purchase,
$p_{CO_2}, p_{CO}, p_{NO_x}\ p_{SO_2}, p_{dust}$	variable in time unit rates for emitting CO_2, CO, NO_x, SO_2, dust, PLN/kg, for example $p_{co_2}(t) = p_{co_2}^{t=0}e^{a_{co_2}t}$,

$\rho_{CO_2}, \rho_{CO}, \rho_{NO_x}, \rho_{SO_2}, \rho_{dust}$	CO_2, CO, NO_x, SO_2, dust emission per fuel chemical energy unit, kg/GJ (depend on fuel type).
r	discount rate,
t_A	annual power plant operation time,
t	time,
T	expressed in years, calculation power plant operation time,
$x_{sw,m,was}$	coefficient taking into account the cost of top-up water, auxiliary materials, wastes.
$x_{sal,t,ins}$	coefficient taking into account the cost of pays, taxes, insurance policies, etc.
z	freezing coefficient [1, 4, 5],

when the number of hydrogen kilograms obtained in the process of electrolysis is expressed by the equation:

$$m_{H_2} = \frac{N_{el}^{RES} t_{RES}}{E_{el}^{H_2}} \tag{5.2}$$

where $E_{el}^{H_2}$ means electric power needed to its generation.

Substituting (5.1) for N_{el}^{RES} and J_{SOE} zero value we obtain the formula for *NPV* profit obtained from power plant operation producing only electricity. The prerequisite for profitable hydrogen production in power units can be presented with the dependency:

$$\begin{aligned} NPV_{block}^{H_2} - NPV = &-N_{el}^{RES} t_{RES} \frac{e_{el}^{t=0}}{a_{el} - r}[e^{(a_{el}-r)T} - 1] + m_{H_2} \frac{e_{H_2}^{t=0}}{a_{H_2} - r}[e^{(a_{H_2}-r)T} - 1] + \\ &- (1 + x_{sal,t,ins}) J_{SOE}(1 - e^{-rT}) \frac{\delta_{serv}}{r} - J_{SOE}\left(\frac{1 - e^{-rT}}{T} + 1\right) \geq 0 \end{aligned} \tag{5.3}$$

For the analysis a *SOE—solid oxide electrolyser* was assumed, where subject to electrolysis is steam collected, for example from a turbine relief valve. *SOE* electrolysers operate at a steam temperature ranging from 650 °C to 900 °C. It should be stressed that a by-product of the electrolysis process is obviously oxygen that can also make a revenue when sold. It would raise the economic effectiveness of hydrogen production process in the power unit. The energy efficiency of the electrolyser is approx. 74%, i.e. the electric energy needed to produce a kilogram of hydrogen equals:

$$E_{el}^{H_2} = \frac{121}{0,74} \frac{\text{MJ}}{\text{kg}_{\text{H}_2}} = 0,045 \frac{\text{MWh}}{\text{kg}_{\text{H}_2}} \tag{5.4}$$

Formula (5.1) shows that the higher is the loss in revenue from selling electric energy, the higher must be the sales price of hydrogen $e_{H_2}^{av}$. The revenue from its sales must at least cover losses caused by lower electricity production, capital costs and repair costs of the electrolyser as well as the cost of employee pays operating it and the cost of insurance policies, etc.

From (5.3) assuming that $NPV_{block}^{H_2} - NPV = 0$ and for $a_{H_2} = 0$ using formula (5.4) we can obtain the dependency for an average unit cost $k_{H_2,block}^{av}$ of hydrogen production in the power unit:

$$k_{H_2,block}^{av} = \frac{E_{el}^{H_2}}{t_{RES}}\left\{(1 + x_{sal,t,ins})i_{SOE}\delta_{serv} + i_{SOE}\left(\frac{r}{T} + \frac{r}{1-e^{-rT}}\right) + t_{RES}\frac{r}{1-e^{-rT}}\frac{e_{el}^{t=0}}{a_{el}-r}[e^{(a_{el}-r)T} - 1]\right\} \tag{5.5}$$

where:

i_{SOE} unit investment outlays on the *SOE* electrolyser (for calculations it was assumed $i_{SOE} = J_{SOE}/N_{el}^{RES} = 8,5$ mln PLN/MW.

The sales price for hydrogen $e_{H_2}^{av}$ generated in the power unit must obviously follow this relation:

$$e_{H_2}^{av} \geq k_{H_2,block}^{av} \tag{5.6}$$

In the case of hydrogen production using electricity generated in *RES* the total discounted profit from its production can be expressed with dependency:

$$NPV_{RES}^{H_2} = \left\{m_{H_2}\frac{e_{H_2}^{t=0}}{a_{H_2}-r}[e^{(a_{H_2}-r)T} - 1] - (1 + x_{sal,t,ins})(J + J_{SOE})(1 - e^{-rT})\frac{\delta_{serv}}{r} + \right.$$
$$\left. -(zJ + J_{SOE})\left(\frac{1-e^{-rT}}{T} + 1\right)\right\}(1-p) \tag{5.7}$$

From which assuming that $NPV_{RES}^{H_2} = 0$ and $a_{H_2} = 0$ we obtain a formula for an average unit production cost of 1 kg of hydrogen:

$$k_{H_2\,RES}^{av} = \frac{E_{el}^{H_2}}{t_{RES}}\left[(1 + x_{sal,t,ins})(i_{RES} + i_{SOE})\delta_{serv} + (zi_{RES} + i_{SOE})\left(\frac{r}{T} + \frac{r}{1-e^{-rT}}\right)\right] \tag{5.8}$$

Equation (5.8) shows that the value of cost $k_{H_2\,RES}^{av}$ using electricity generated in the wind turbine set is significantly lower than the same cost in a photovoltaic plant. The time of wind turbine set operation is significantly higher than the time of the photovoltaic plant. Under Polish conditions $t_{RES}^{turb} \cong 1750$ h/rok, $t_{RES}^{photo} \cong 750$ h/rok($i_{RES} \approx 6.5$ million PLN/MW, Table 5.1).

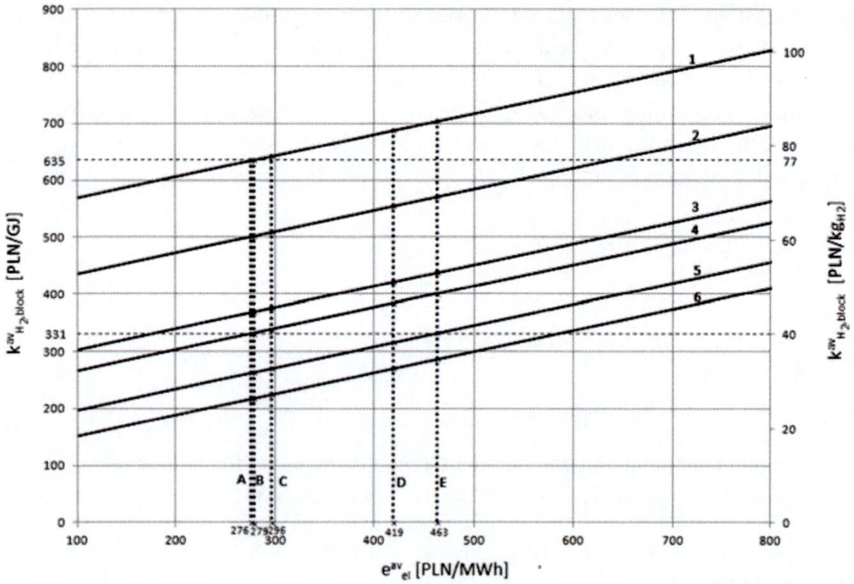

Fig. 5.3 The average hydrogen production cost $k^{av}_{H_2,block}$ for the power unit as the function of electricity price for different operation times of *RES*, where: 1—t_{RES} = 750 h/rok, 2—t_{RES} = 1000 h/rok, 3—t_{RES} = 1500 h/rok, 4—t_{RES} = 1750 h/rok, 5—t_{RES} = 2500 h/rok, 6—t_{RES} = 3500 h/rok(A—unit electricity production cost in the hierarchical gas–gas and gas-steam technology, B—in steam technology for supercritical parameters with burning coal in air atmosphere, C—in the technology of the dual-fuel gas-steam in a parallel system, D—in nuclear technology, E—in steam technology for supercritical parameters with burning coal in oxygen atmosphere)

5.2.1 *Unit Hydrogen Production Cost*

Figure 5.3 presents calculation results of unit production cost $k^{av}_{H_2,block}$ of hydrogen in the power unit, (formula (5.5)), as the function of average integral sales price e^{av}_{el} of electricity generated in it. The variable parameter in calculations is the time of *RES* operation, when the power unit produces not only electricity, but also hydrogen in the electrolyser.

The mean integrated price of sales of electricity produced in the power unit is expressed by the following equation:

$$e^{av}_{el} = \frac{1}{T}\int_0^T e^{t=0}_{el} \mathrm{e}^{a_{el}t}dt = \frac{e^{t=0}_{el}}{Ta_{el}}\left(\mathrm{e}^{a_{el}T} - 1\right). \quad (5.9)$$

Dotted vertical lines A, B, C, D, E in Fig. 5.3 correspond to the cost from Table 5.2, of unit costs $k_{el,av}$ for the production of electricity in the power unit for various power generation technologies. The corresponding values $k^{av}_{H_2,block}$ are minimum values for this technology. As they correspond to the situation, when the sales price e_{el}^{av} equals $k_{el,av}$. For example for $t_{RES} = 750$ h/year and $k_{el,av} = 276$ PLN/MWh the unit hydrogen production cost in the hierarchical *gas–gas* and *gas-steam* technology $k_{H_2,block}^{av} = 635$ PLN/GJ (77 PLN/kg_{H2}). This cost is many times higher than the price of very expensive Russian natural gas amounting in Poland to 32 PLN/GJ. In practice, power plants have to generate profit, so the price e_{el}^{av} must be higher than $k_{el,av}$. Here, it should be emphasised that the most economically viable would be to produce hydrogen in nuclear power stations, where the cost of electricity production can be lower than 200 PLN/MWh—Fig. 5.2. The production cost for hydrogen in those stations would be the lowest, lower than 200 PLN/GJ (25 PLN/kg_{H2})—Fig. 5.4.

Figure 5.4 presents the calculation results for the unit cost of hydrogen production using electricity generated in *RES* as the function of its operation time. For example for $t_{RES} = 750$ h/year, i.e. for the average operation time of the photovoltaic plant in Poland the unit cost of hydrogen production using the electricity generated in that plant is as high as $k_{H_2,block}^{av} = 957$ PLN/GJ (116 PLN/kg_{H2}). The production of hydrogen using electric energy from the power plant is cheaper by a few dozen percent—Fig. 5.3.

It should have been expected, because the lower electricity production costs, the lower costs of hydrogen production. However, it does not change the fact that costs of hydrogen produced by means of electricity generated in the power plant using any technology as well as in *RES* are high. They exceed the price of hydrocarbon fuels many times.

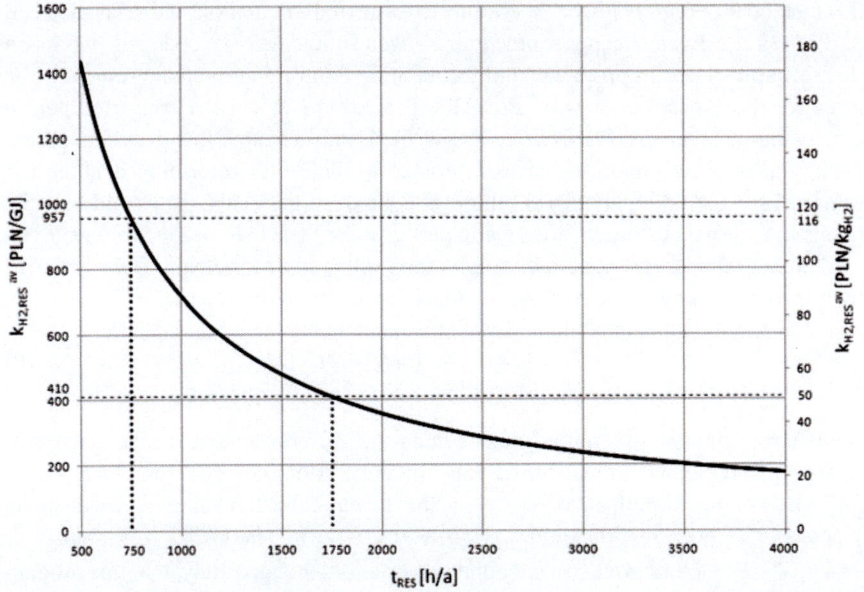

Fig. 5.4 The average hydrogen production cost using electricity generated in *RES* (the photovoltaic plant and wind turbine set) as the function of their operation

5.3 Summary

The analyses results presented in the chapter show that the cost of hydrogen production using electricity, especially from *RES* is high. Here, it should be emphasised, which was already done, that the most economically viable would be to produce hydrogen in nuclear power stations, where the cost of electricity production can be lower than 200 PLN/MWh—Fig. 5.2.

The fundamental disadvantage of hydrogen production using the process of water electrolysis is the fact that from the amount of approx. 180 MJ electric power, the most noble energy form, we obtain only one kilogram of hydrogen (its calorific value is $LHV = 121\ \mathrm{MJ}/\mathrm{kg_{H_2}}$) that can return only about 60 MJ of electricity. Therefore, this production is a "thermodynamic barbarity". Currently, hydrogen is obtained mainly in the process of natural gas reforming with water steam according to an endothermic reaction: $CH_4 + H_2O \rightarrow CO + 3H_2$ (in Poland this technology is used to produce about 1 million tonnes of hydrogen a year). The energetic demand for heat (heat features, as opposed to electricity, low quality, i.e. low exergy) for the reaction amounting to 207 MJ/kmol CH_4. Therefore, they are above 5 times lower per 1 kg of hydrogen obtained than energetic needs in the process of water electrolysis, where it is, additionally, supplied not by heat, but electricity. The energy identified with exergy, that is to say the energy with the highest thermodynamic quality, so expensive, while heat features low quality, i.e. low exergy, and is relatively cheap. In addition, investment outlays on reforming plants are small as compared to outlays on electrolysers and electricity sources. Thus, the unit cost of hydrogen obtained using this method is definitely lower. According to estimated calculations it does not exceed $13\ \mathrm{PLN}/\mathrm{kg_{H_2}}$. Even cheaper source of hydrogen should be coke oven gas (works on this technology are in progress) and gas coming from mine methane removal (the price for those gases is approx. PLN 200 per 1,000 Nm^3; when converted per an energy unit is approx. 8 PLN/GJ; i.e. gases are 4 times cheaper than Russian natural gas). Annual total amount of gases in Poland available for reforming is about 2.5 billion Nm^3, including about 1.5 billion of coke oven gas. This gas is sold by coke plants to external customers after using part of it for their own production needs.

Another, almost the same barbarity is the production of methane from obtained hydrogen according to the exothermic reaction:

$$CO_2 + 4H_2 \rightarrow CH_4 + 2H_2O \tag{5.10}$$

(very large amount of emitted heat equals the difference in calorific values of hydrogen and methane; the calorific value of a one kilomole hydrogen is $LHV_{H_2} = 242\ \mathrm{MJ}/\mathrm{kmol_{H_2}}$; $\mathrm{kmol_{H_2}} = 2\ \mathrm{kg}_{H_2}$; the lower calorific value of methane is $LHV_{CH_4} = 802,32\ \mathrm{MJ}/\mathrm{kmol_{CH_4}}$, $LHV_{CH_4} = 50,15\ \mathrm{MJ}/\mathrm{kg_{CH_4}}$, $\mathrm{kmol_{CH_4}} = 16\ \mathrm{kg}_{CH_4}$). Sometimes one can encounter an opinion, though rarely, that this production binds carbon dioxide generated from firing coal, i.e. it automatically eliminates its problem. It shows a complete lack of understanding for the occurring thermodynamic reactions and phenomena. To get rid of CO_2, we "destroy" electricity generated in the power plant reducing it from (see formula (5.9)) 1440 MJ $= 4\ \text{kmoles}_{H_2} \times 2\ \text{kg}_{H_2}/\text{kmol}_{H_2} \times 180\ \text{MJ}/\text{kg}_{H_2}$ to 400 MJ, i.e. almost in 75% (from one kilomole of methane one can obtain approx 400 MJ of electricity). The same nearly could be obtained, if the electricity was not generated at all. Furthermore, then it would not be necessary to build any power plant. So a sort of technical "juggling act" using lots of money spent on electricity destroying plants in the service of the lie concerning the supposed global warming caused by CO_2 emission and resulting from it a supposed necessity to build *RES* (it should be emphasised that the water steam generated from burning hydrogen is far more a greenhouse gas than CO_2). At the same time, *RES* are not capable of producing electricity to compensate the effects of its "destruction" in power plants. To make up for the losses generated by *RES* and regain the 1040 MJ (= 1440 –400) of electric power, the coal consumption has to be doubled! It should be complemented with millions of tonnes of coal used during hundreds of power unit starts after their shutoffs as a result of renewable energy source *RES* operation. Unfortunately, when starting power units, the electricity is not delivered to the grid. Is this all consistent with the supposed necessary power industry decarbonisation programme, allegedly for the benefit of mankind? Furthermore, the carbon dioxide emission by European countries makes up only about 5% of global emission from burning fossil fuels. This global fuel emission makes up only 3.3% of global CO_2 emission (5% of European power industry is responsible for only $0.16\% = 0.05 \times 0.033$ of global emission, which is almost zero!). Then, where is the rest? The oceans produce 41.4% of carbon dioxide, while the biosphere 55.3% (for example humans breathe out daily 1 kg of CO_2, during physical stress 4 kg CO_2). Therefore, even a complete shutdown of not only Polish, but all global fossil fuel-fired power plants (obviously without nuclear power stations) can hardly change anything. What is more, as humans need oxygen to live, the biosphere, apart from living organisms, needs carbon dioxide to exist.

References

1. Bartnik R, Buryn Z (2011) Conversion of coal-fired power plants to cogeneration and combined-cycle: thermal and economic effectiveness. Springer, London
2. KowalczyK T, Ziółkowski P, Badur J (2015) Exergy losses in the Szewalski binary vapor cycle. Entropy 17:7242–7265. https://doi.org/10.3390/e17107242
3. Bartnik R, Buryn Z, Hnydiuk-Stefan A (2017) Methodology and a continuous time mathematical model for selecting the optimum capacity of a heat accumulator integrated with a CHP plant (in Polish: Ekonomika Energetyki w modelach matematycznych z czasem ciągłym). Wydawnictwo Naukowe PWN, Warszawa
4. Bartnik R (2009) Gas-steam power plants and combined heat and power plants. Energy and economic efficiency (in Polish: Elektrownie i elektrociepłownie gazowoparowe. Efektywność energetyczna i ekonomiczna), WNT, Warszawa (reprint 2012, 2017).
5. Bartnik R, Bartnik B (2014) Economic calculation in the energy sector (in Polish: Rachunek ekonomiczny w energetyce). WNT, Warszawa

6. Feidt M (2018) Finite Physical Dimensions Optimal Thermodynamics 2-Complex Systems, ISTE Press–Elsevier, 2018
7. Sieniutycz S., Jezowski J., Energy Optimization in Process Systems, Elsevier Science, 2009

Chapter 6
Thermodynamic and Economic Analysis of a Hierarchical Gas-Gas Engine Integrated with a Compressed Air Storage

6.1 Introduction

Natural hollow spaces in the rock mass (caverns) or underground mining excavations can be used as compressed air storages with which electricity can be stored (a significant problem may be the leakage of these spaces; another well established way of storing electricity is to store it with the energy of potential water in the upper reservoirs of pumped hydroelectric power plants). The necessity to store electricity results, among others, from from the need to defend fossil fuel-based electricity sources against the so-called renewable sources (RES), i.e. against wind turbines and photovoltaic cells. These sources enjoy priority access to the power grid supplying electricity to consumers, which means that fossil fuel sources have to adapt to their very "chimeric" operation. As a result, conventional sources are shut down hundreds of times a year or rapidly reduce their power, which has a very negative impact on their technical lifetime. They degrade quickly which force them to undergo frequent and costly overhauls. To avoid this, it is necessary to ensure their stable operation by storing the electricity they produce, for example, using compressed air in caverns.

Figure 6.1 shows a schematic diagram of this storage method. The electric motor of the compressor C_M charges the air storage during off-peak hours at the power of ΔN_a (Fig. 6.3). During peak hours, the compressor C_M is shut down and the inlet guide vanes of the compressor C_{GT} are closed and air of pressure p_1^{opt} (Fig. 6.5) is fed to the CH combustion chamber of the gas turbine directly from the compressed air storage after being heated to the temperature of T_{10} in the heat exchanger W—Fig. 6.1.

Heating the air from the cavern from temperature T_{cav} to temperature T_{10} is necessary so that the low value of T_{cav} results in as little reduction as possible in the turbine efficiency $\eta_{GT}^{\max}$, and thus in the engine efficiency η_{G-G}. This negative effect of a low value of T_{cav} on efficiency η_{G-G} is the greater the higher the temperature T_2. This is due to the fact that with the T_2 increasing, the difference between T_1 and T_8 increases (Figs. 3.7 and 4.21), and therefore obviously between T_1 and T_{10}. In economic calculations, the temperature reduction from T_1 to T_{10} can be taken into account by reducing the z_{GT}^{opt} in formulas (6.35) and (6.36) (formulas (4.8), (6.8)) or

R. Bartnik and T. W. Kowalczyk, *Hierarchical Gas-Gas Systems*,
Power Systems, https://doi.org/10.1007/978-3-030-69205-6_6

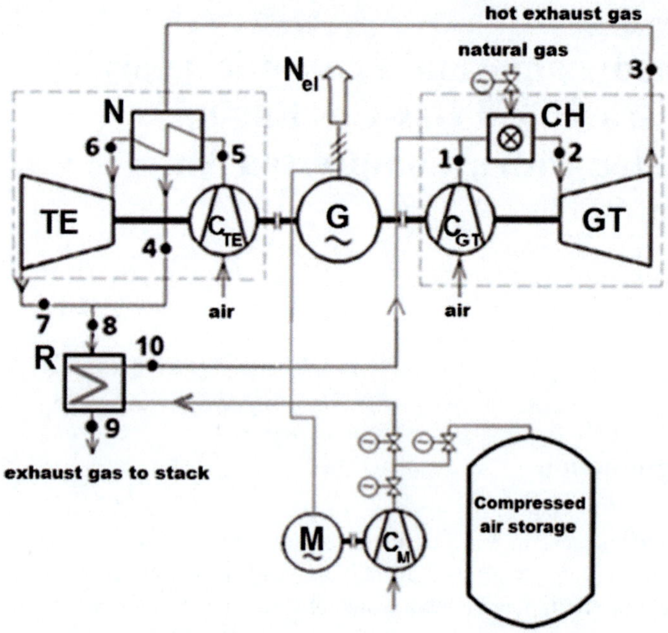

Fig. 6.1 Schematic diagram of a hierarchical gas-gas engine working together with a compressed air storage

by adopting a relatively large capital expediture J_{CAES} on the compressed air storage infrastructure, as was done in this chapter.

The rectangles drawn in Fig. 6.1 mean that the turboexpander TE together with the low-pressure compressor C_{TE} are built in one housing and on a common shaft, just like it is done in practice for the TG turbine part and the high-pressure compressor C_{GT} of the gas turbine. This reduces the capital expediture on in the engine.

Chapter 5 describes the storage of electricity with the use of hydrogen obtained in the electrolysis of water. However, this method is highly unreasonable, both thermodynamically and economically. It should even be said expressis verbis that it is absurd, that it is thermodynamic and economic "barbarism". It literally "destroys" electricity, the noblest, most valuable form of energy, in every sense of these words, generated in fossil fuel sources, the "destroyed" energy of which cannot be produced by renewable energy sources. Moreover, from the obtained kilogram of hydrogen with a lower heating value of $LHV = 121\ MJ/kg_{H_2}$ only around 60 MJ of electricity can be recovered, i.e. only 33% of the electricity used to produce it (60 MJ = 0.33 × 180 MJ). The remaining 67% is therefore irretrievably lost (sic). Not only that, the energy produced in RES is many times more expensive than energy from sources using fossil fuels and therefore requires huge subsidies from state budgets. Therefore, RES are only a source of unjustified huge annual financial benefits for their owners at the expense of taxpayers. In this light, it is clear that using compressed air for electricity storage compared to hydrogen storage is a thermodynamically and economically rational way. This chapter therefore analyzes this method.

6.2 Thermodynamic Analysis of a Hierarchical Gas-Gas Engine Cooperating with a Compressed Air Storage Facility

The analysis was carried out for two variants of the operation of a hierarchical gas-gas engine cooperating with a compressed air storage. One for the engine in which a *Joule-Brayton* cycle is realized in the gas turbine with one-stage compression of the circulating medium, Fig. 4.3, and the other with a its two-stage compression process and inter-stage cooling, Fig. 6.2, which is obviously thermodynamically advantageous. On the other hand, in both variants a *Joule-Brayton* cycle of the turboexpander operating in the low temperature range is a cycle with only one-stage compression. It is not profitable to improve the turboexpander cycle thermodynamically due to its relatively low power compared to the power of the gas turbine—Fig. 6.4. Such an improvement would not bring a significant increase in its power, and thus would not bring tangible economic benefits, but would only unnecessarily increase the capital expediture on the engine (see Sect. 4.3).

In the calculations for the gas-gas engine variant with the *Joule-Brayton* cycle of a gas turbine with two-stage compression—Fig. 6.2—the temperature T_4 at the inter-cooler exit was assumed to be 20 °C higher than the ambient temperature T_{amb} (obviously, the closer T_4 temperature gets to T_{amb}, the closer the adiabatic compression power will be to the thermodynamic ideal, i.e. the minimum isothermal compression power). The minimum adiabatic compression power $N_{i\ C}$ (formula (6.1)), assuming

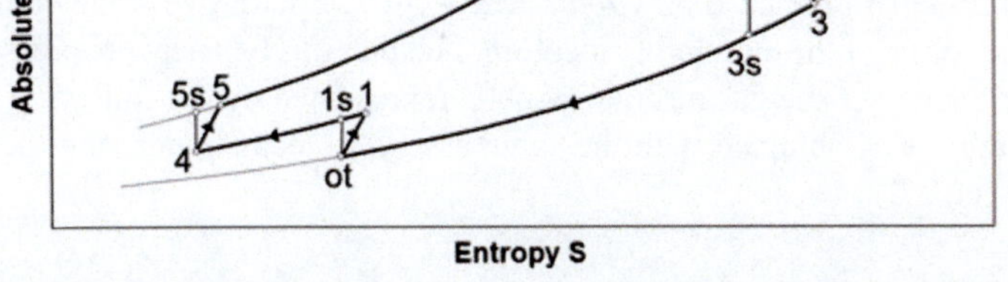

Fig. 6.2 The *Joule-Brayton* cycle of a gas turbine with two-stage compression and inter-stage cooling of the circulating medium

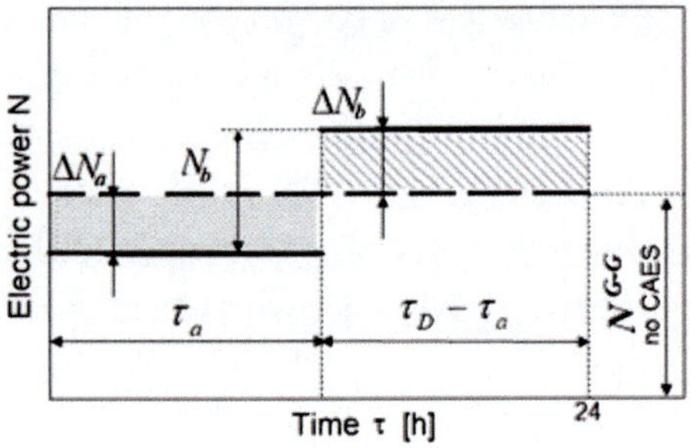

Fig. 6.3 The *gas-gas* engine load with and without compressed air accumulator throughout the day (τ_D —number of hours in a day, τ_a —number of off-peak hours)

the same internal efficiency values η_i^C, in both stages is guaranteed by the equality $T_{5s} = T_{1s}$ [1].

Figure 6.3 shows the time diagram of the daily operation of the hierarchical gas-gas engine cooperating with the compressed air accumulator. The bold dashed line in this figure also shows the engine power N^{G-G} it would have worked with were there no accumulator in the system.

During time τ_a of reduced demand for electricity in the power grid, electricity is returned from the *gas-gas* engine of capacity N^{G-G} in amount of $(N^{G-G} - \Delta N_a)\tau_a$; during peak hours $\tau_b = \tau_D - \tau_a$—in the amount of $(N^{G-G} + \Delta N_b)\tau_b$. If the engine does not cooperate with the compressed air storage, the energy returned to the network amounts to $N^{G-G}\tau_D$. For each mode of the daily operation of the system, the amount of the achieved profit is different. It can be expected that due to the higher price of electricity at the peak of demand than the price during off-peak hours the operation with variable power will be more economically advantageous, despite the fact that the system with a compressed air storage system is more expensive in terms of investment. In the economic analysis it was assumed that the increases in engine power in the peak hours and in the off-peak hours are equal $\Delta N_a = \Delta N_b$ and equal to the power $N_{i\,C}$ of compressing the air in a gas turbine compressor C_{GT}. Of course, for thermodynamic reasons it would be most advantageous if the power of $N_{i\,C}$ was as low as possible, because then - for a given power of expansion $N_{i\ exp}$ of the medium in the turbine - the power of the gas turbine $N^{GT} = N_{i\ \mathrm{exp}} - N_{i\,C}$ would be greater (powers $N_{and\ S}$ i $N_{i\ exp}$ result from the energy balance of the *Joule-Brayton* cycle of the gas turbine). In the case of a *Joule-Brayton* cycle with two-stage compression with intercooling of the circulating medium, Fig. 6.2, this balance—expressed by the energy efficiency of the gas turbine η_{GT}—is represented by the following equation (this equation was obtained with the same assumptions as in deriving the formula (3.9) in Sect. 3.2):

$$\eta_{GT} = \frac{N^{GT}}{\dot{E}_{ch}} = \frac{N_{i\ \mathrm{exp}} - N_{i\,C}}{\dot{E}_{ch}} = \frac{\eta_m(T_2 - T_3) - \left[\frac{1}{\eta_m}(T_1 - T_{amb}) + \frac{1}{\eta_m}(T_5 - T_4)\right]}{T_2 - T_5} =$$

$$= \frac{\eta_m \eta_i^{GT} T_2\left(1 - \frac{1}{z_{GT}^2}\right) - \frac{T_{amb}+T_4}{\eta_m \eta_i^C}(z_{GT} - 1)}{T_2 - T_4(1 + \frac{z_{GT}-1}{\eta_i^C})} \rightarrow \max \tag{6.1}$$

where the final form of the formula (6.1) is obtained by substituting the temperatures resulting from the irreversible adiabates ot-1, 4–5, 2–3 (Fig. 6.2):

$$T_1 = T_{amb} + \frac{1}{\eta_i^C}(T_{1s} - T_{amb}), \tag{6.2}$$

$$T_5 = T_4 + \frac{1}{\eta_i^C}(T_{5s} - T_4) \tag{6.3}$$

$$T_3 = T_2 - \eta_i^{GT}(T_2 - T_{3s}) \tag{6.4}$$

and the relationships of an isentropic process:

$$z_{GT} = \frac{T_{1s}}{T_{amb}} = \frac{T_{5s}}{T_4} = \left(\frac{p_1}{p_{amb}}\right)^{\frac{\kappa-1}{\kappa}} = \left(\frac{p_5}{p_4}\right)^{\frac{\kappa-1}{\kappa}}. \tag{6.5}$$

and the relationships of an isentropic process:

$$z_{GT}^2 = \frac{T_2}{T_{3s}} = \left(\frac{p_2}{p_{amb}}\right)^{\frac{\kappa-1}{\kappa}} = \left(\frac{p_5}{p_4}\frac{p_1}{p_{amb}}\right)^{\frac{\kappa-1}{\kappa}}. \tag{6.6}$$

The optimum value z_{TG}^{opt}, i.e. the value guaranteeing the maximum energy efficiency $\eta_{TG}^{\max}$ of the gas turbine circuit (formula (6.1)), results from the condition:

$$\frac{d\eta_{GT}}{dz_{GT}} = 0. \tag{6.7}$$

Differentiation yields the following equation, used to calculate z_{GT}^{opt}:

$$\left[2a - b(z_{GT}^{opt})^3\right]\left[T_2 - T_4\left(1 + \frac{z_{GT}^{opt} - 1}{\eta_i^C}\right)\right] - \frac{T_4}{\eta_i^C}\left[a(z_{GT}^{opt})^3 - az_{GT}^{opt} - b(z_{GT}^{opt})^4 + b(z_{GT}^{opt})^3\right] = 0 \tag{6.8}$$

where:

$$a = \eta_m \eta_i^{GT} T_2,\; b = \frac{T_{amb} + T_4}{\eta_m \eta_i^C} \tag{6.9}$$

Using z_{GT}^{opt} the following dependency is obtained:

$$\frac{N_{i\ C}}{N_{i\ \exp}} = \frac{\frac{1}{\eta_m}(T_1 - T_{amb}) + \frac{1}{\eta_m}(T_5 - T_4)}{\eta_m(T_2 - T_3)} = \frac{N_{i\ C}}{N^{GT} + N_{i\ C}} = \frac{(z_{GT}^{opt})^2}{\eta_m^2 \eta_i^C \eta_i^{GT}(z_{GT}^{opt} + 1)} \frac{T_{amb} + T_4}{T_2} \tag{6.10}$$

from which the relationship between the powers z której otrzymuje się ostatecznie związek pomiędzy mocami $N_{i\ C}$ and N^{GT} is ultimately obtained and which, very importantly, is needed in economic analysis (formula (6.36))

$$N_{i\ C} = N^{GT} \frac{(z_{GT}^{opt})^2 (T_{amb} + T_4)}{\eta_m^2 \eta_i^C \eta_i^{GT} T_2 (z_{GT}^{opt} + 1) - (z_{GT}^{opt})^2 (T_{amb} + T_4)} \tag{6.11}$$

Equation (6.11) is very important because it generalises the economic viability of using compressed air storage for any gas turbine power N^{GT} (formula (6.36)). The second basic quantity characterizing the turbine is the temperature of T_2 of the exhaust gases that come from the CH combustion chamber—Fig. 3.2 and 4.1 (the temperature of T_2 is always given in catalogues by manufacturers; the higher, obviously, the higher is the efficiency of $\eta_{GT}^{\max}$). Therefore, the "economic" formulas (6.35), (6.36) expressed in these two technical characteristics are generic, in other words, they apply to each turbine produced. Therefore, the results of the calculations obtained by means of these quantities and presented in the graphs are universal. Their applicability cannot therefore be overestimated.

Analogously to the formula (6.11), the equation for the relationship between the power of $N_{i\ C}$ and N^{GT} for *Joule-Brayton* cycle with single-stage compression of the circulation medium is obtained (Sect. 4.2.1, Fig. 4.3):

$$N_{i\ C} = N^{GT} \frac{z_{GT}^{opt} T_{amb}}{\eta_m^2 \eta_i^C \eta_i^{GT} T_2 - z_{GT}^{opt} T_{amb}}. \tag{6.12}$$

It should be noted at this point that only single-stage compression gas turbines are actually produced, so the results of the economic calculations presented in Sect. 6.3 have practical value only for them. By all means, this does not detract from the results presented for the *Joule-Brayton* cycle with a two-stage compression. This is because it shows, what is of a very important cognitive value, that in spite of the fact that it is thermodynamically more advantageous, Fig. 6.4, its use in the system of a gas engine with a compressed air accumulator would be economically less profitable (Sect. 6.3, Fig. 6.8), precisely because of this greater thermodynamic efficiency. This is because less compression power $N_{i\ C}$ is then needed, so the revenue from peak electricity sales would be lower. It is therefore not always the case that higher thermodynamic efficiency translates into higher cost-effectiveness. This is especially true when technical processes are being interfered with by politicians who are not familiar with them. The thermodynamic benefits of a two-stage compression cycle would be even more significant, were they additionally accompanied by two-stage expansion with intercooling and heat regeneration. This is because it is much closer to the thermodynamic ideal of the Carnot cycle (Fig. 6.15), and thus even less power

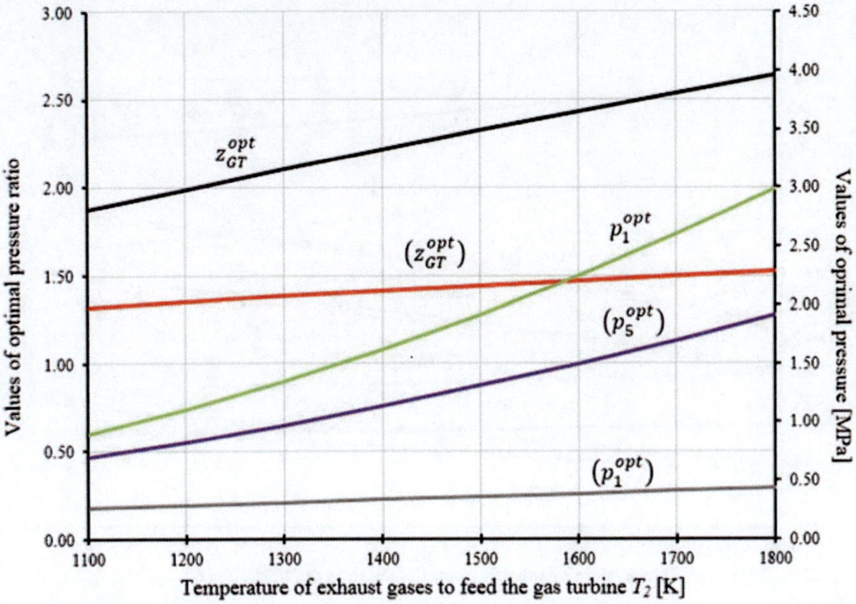

Fig. 6.4 Power ratios $N_{i\ S}/N_{i\ \exp}$, N^{TE}/N^{TG} and efficiencies η_{G-G}, $\eta_{TE}^{\max}$, $\eta_{TG}^{\max}$ as a function of temperature T_2 for *Joule-Brayton* cycle of a gas turbine with one- and two-stage compression (values in brackets refer to *Joule-Brayton* cycle with two-stage compression)

$N_{i\ C}$ is needed. However, this cycle would be economically unviable to the highest degree. Thus, the unreasonable requirement imposed by politicians to build and use renewable energy sources (RES) and granting them priority access to the power grid unfortunately excludes the use of thermodynamically beneficial technical solutions. It can also be expected that economically advantageous despite the increased expenditure on the "better" engine itself. It should be noted that significant capital expenses on the entire electricity storage infrastructure would then be unnecessary. Therefore, if it were not for politicians, it would be by all means most thermodynamically and economically advantageous to dispense with electricity storage and generate electricity in off-peak hours in high-efficiency thermodynamic cycles with a reduced capacity of a gas turbine compared to its nominal capacity (i.e. the capacity with which it would work at the peak of the grid load), despite the fact that operation with reduced capacity also takes place with reduced energy efficiency. The displacement of better solutions by worse is unfortunately a great, timeless truth. In economics, for example, Copernicus-Gresham's law has existed for millennia: bad money drives out good.

Figures 6.4, 6.5 and 6.6 show the results of multivariate thermodynamic calculations for *Joule-Brayton's* gas turbine cycle with one-stage and two-stage compression.

As can be seen from the calculation results presented in Fig. 6.4, the lower the temperature T_2 of the inlet exhaust gas from the combustion chamber to the turbine,

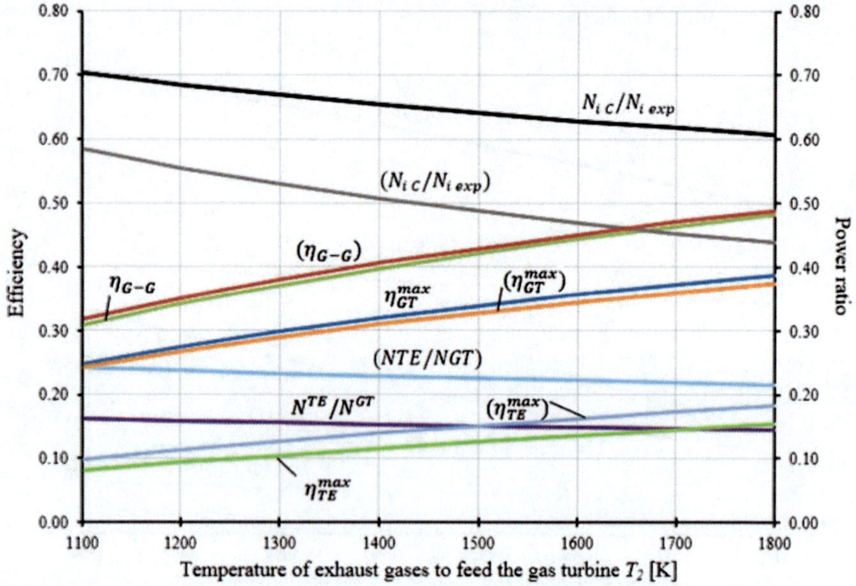

Fig. 6.5 Values of optimal pressures p_1^{opt}, p_5^{opt} and optimal compression ratio z_{GT}^{opt} as a function of temperature T_2 for *Joule-Brayton* cycle of a gas turbine with one- and two-stage compression (values in brackets refer to *Joule-Brayton* cycle with two-stage compression)

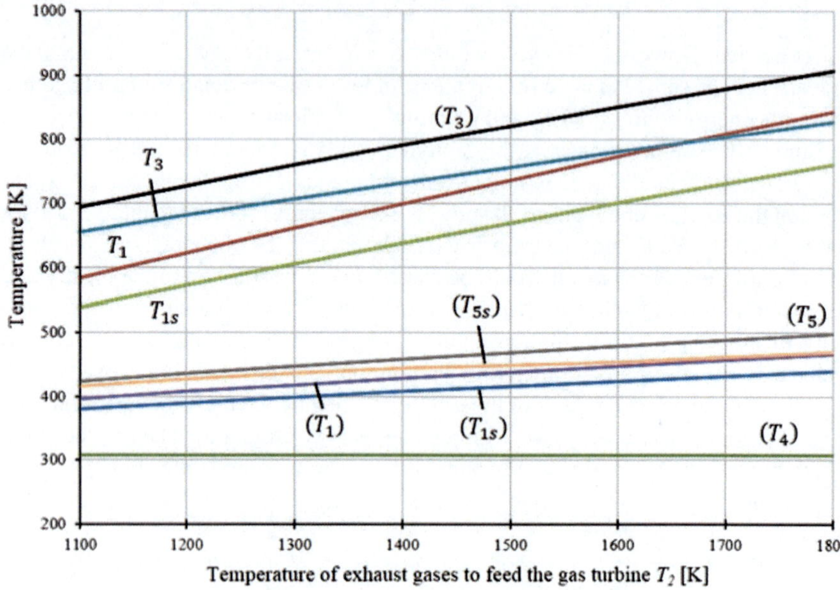

Fig. 6.6 Values of temperatures $T_1, T_{1s}, T_3, T_4, T_5, T_{5s}$ as a function of temperature T_2 for Joule cycle of a gas turbine with one- and two-stage compression (values in brackets refer to *Joule-Brayton* cycle with two-stage compression)

the greater the compression power $N_{i\,C}$. The high power $N_{i\,C}$ is obviously thermodynamically unfavorable, but when compressed air accumulators are used for electricity storage, Fig. 6.1, it is economically advantageous. The results of economic calculations presented in the paper were made for the temperature $T_2 = 1800\ K$—Figs. 6.9, 6.11, 6.12, 6.14 (only Fig. 6.13 was made for the temperature $T_2 = 1400\ K$ to show the increase in profit NPV_{CAES} with decreasing temperature T_2, see formula (6.39)). This was done because for the temperature $T_2 = 1800\ K$ (it is currently the highest permissible temperature of the exhaust gas fed to the turbine due to the limited creep resistance of its blades) the economic efficiency is the lowest, as the peak power $N^p = 2N_{i\,C}$ is the lowes (Fig. 6.3), which results in the lowest revenues from electricity peak sales. Thus, the profit NPV_{CAES} (formulas (6.35), (6.36)) presented in Fig. 6.9 is "safely underestimated", because it should be expected that in practice it will be higher due to the actually lower temperatures T_2.

As already mentioned above, the minimum adiabatic compression power $N_{i\,C}$ (formula (6.1)) assuming the same internal efficiency values η_i^C in both stages is guaranteed by the equality $T_{5s} = T_{1s}$. For the value of z_{GT}^{opt}, which in turn guarantees the maximum efficiency of η_{GT}^{max}, these temperatures are different—Fig. 6.6.

6.2.1 *Minimum Required Volume of the Compressed Air Storage*

To determine the minimum required volume of the compressed air storage, i.e. the volume that guarantees that the pressure in it does not fall below the permissible pressure value in the combustion chamber of a gas turbine, the thermal equation of the gas state should be used. In this chapter the *Clapeyron* equation for perfect and semi-perfect gases is used. The application of "mathematically complicated" equations for real gases (e.g. the virial equation of state) will not cause the storage volume values obtained from them to differ significantly from those obtained from the *Clapeyron* equation. Therefore their use is not justified, it would only "blur" unnecessarily the "simplicity" of the calculation of the storage volume (formula (6.15)).

From the *Clapeyron* equation for a compressed air storage of a volume $V_{cav}^{\min}$ before it is discharged (index p):

$$p_p V_{cav}^{\min} = m_p R_{air} T_{cav} \tag{6.13}$$

and after it is discharged (index k):

$$p_k V_{cav}^{\min} = m_k R_{air} T_{cav} \tag{6.14}$$

the formula for the minimum required volume of the cavern is obtained, which will ensure that during peak hours it will be possible to close the inlet guide vanes of the compressor S_{TG} (Figs. 3.1 and 4.1) and to feed the pressurized air at pressure $p_p = p_1^{opt}$ from the cavern (Fig. 4.22) to the turbine combustion chamber:

$$V_{cav}^{\min} = \frac{\dot{m}_a(\tau_D - \tau_a)}{\Delta p_{cav}} R_{air} T_{cav} \tag{6.15}$$

with the air intake from the cavern at peak demand $\tau_b = \tau_D - \tau_a$ (Fig. 6.2) being determined from the equation:

$$\dot{m}_a(\tau_D - \tau_a) = m_p - m_k \tag{6.16}$$

and the allowable pressure drop in it is:

$$\Delta p_{kaw} = p_p - p_k \tag{6.17}$$

where:

$\dot{m}_a$ flow rate of the air supplied to the gas turbine combustion chamber (obviously, the greater the power of the gas turbine, the greatervalue $\dot{m}_a$),

R_{air} air gas constant, $R_{air} = 287{,}04$ J/(kg K),

T_{cav} air temperature in the cavern (it was assumed that the temperature of T_{kaw} is constant).

In order for the air pressure at the inlet to the KS combustion chamber of the turbine to be constant and equal to p_1^{opt}, the pressure p_p in the reservoir should be greater than p_1^{opt} and during discharging it should be throttled to to p_1^{opt} by means of a control valve installed at the exit of the reservoir. The isenthalpic throttling that takes place in the valve is, however, a source of exergy losses.

Using the formula (6.15), multi-variant calculations of the volume V_{kaw}^{min} were performed, the results of which are presented in Fig. 6.7.

It is clear that the greater the allowable pressure drop Δp_{cav}, the smaller the volume of the cavern $V_{cav}^{\min}$ can be, and in turn the longer the time τ_b and the higher the air flow rate $\dot{m}_a$, the larger the volume must be.

Figure 6.8 shows the relative reduction in the efficiency η_{G-G} of the gas engine by a value $\Delta\eta_{G-G}$ due to the pressure drop Δp_{cav} in the cavern.

As shown in Fig. 6.8, the pressure drop Δp_{kaw} has only a negligible influence on the efficiency of the engine η_{GG}. Besides, as already mentioned above, in order to completely eliminate the influence of Δp_{cav} on the efficiency of η_{G-G}, the pressure of p_p in the tank should be greater than p_1^{opt} and it should be throttled when supplying air to the combustion chamber CH.

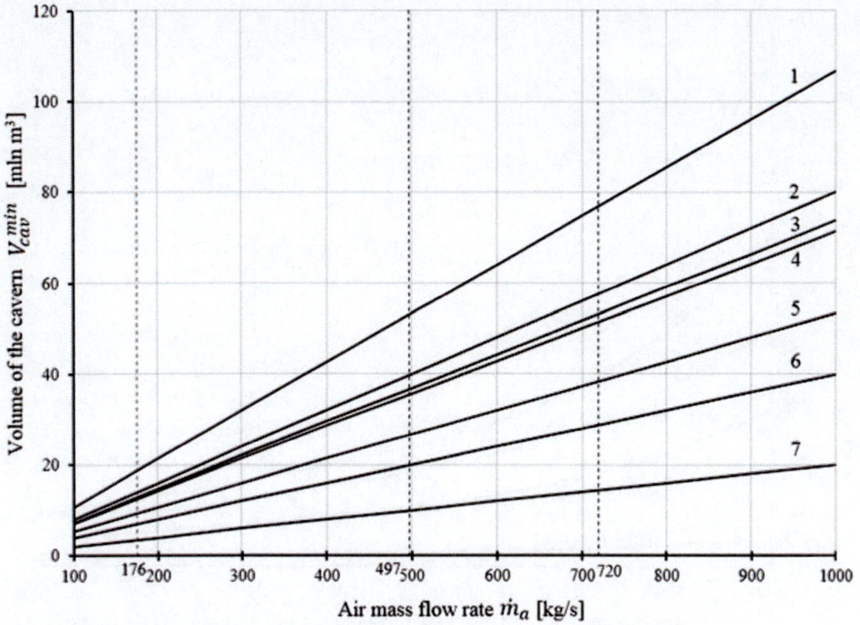

Fig. 6.7 Required minimum volume of the compressed air storage $V_{kaw}^{\min}$ as a function of the air flow rate $\dot{m}_a$ with Δp_{cav}, τ_b, as parameters, where: 1—$\Delta p_{cav} = 0{,}5$ bar, $\tau_b = 12$ h, $T_{cav} = 323$ K, 2 —$\Delta p_{cav} = 0{,}5$ bar, $\tau_b = 16$ h, $T_{cav} = 323$ K, 3—$\Delta p_{cav} = 0{,}5$ bar, $\tau_b = 12$ h, $T_{cav} = 298$ K, 4—$\Delta p_{cav} = 0{,}5$ bar, $\tau_b = 12$ h, $T_{cav} = 288$ K, 5—$\Delta p_{cav} = 0{,}5$ bar, $\tau_b = 8$ h, $T_{cav} = 323$ K, 6—$\Delta p_{cav} = 1{,}0$ bar, $\tau_b = 12$ h, $T_{cav} = 323$ K, 7—$\Delta p_{cav} = 2{,}0$ bar, $\tau_b = 12$ h, $T_{cav} = 323$ K (air flow rates $\dot{m}_a = 176$ kg/s, 497 kg/s and 720 kg/s correspond to 52.8 MW GE GT8C, 202 MW Siemens SGT6-5000F and 334 MW MHPS M701G)

6.3 Economic Analysis of the Use of Compressed Air Storage as a Way of Storing Electricity

For the economic analysis of the hierarchical cooperation of a *gas-gas* engine with a compressed air accumulator, it is best to use an innovative methodology of recording economic measures of profitability of any business undertaking in a continuous time [2–4]. This methodology allows to obtain a continuous function of discounted profit $NPV_{CAES} = f(N^{TG})$. This is very important because the knowledge of its course in the whole range of gas turbine power variability $N^{TG} \in \langle 0; \infty)$ (a gas turbine thermodynamically characterizes the entire *gas-gas* engine) gives, which should be emphasized with all its might, a comprehensive look at the problem under consideration. The known and so far used discrete record of the NPV discounted profit meter [5] does not give such a possibility. Calculating a few or even several dozen NPV values using it does not allow to assess the character of the NPV curve. Thus the "point" results do not give the possibility of drawing conclusions of general nature,

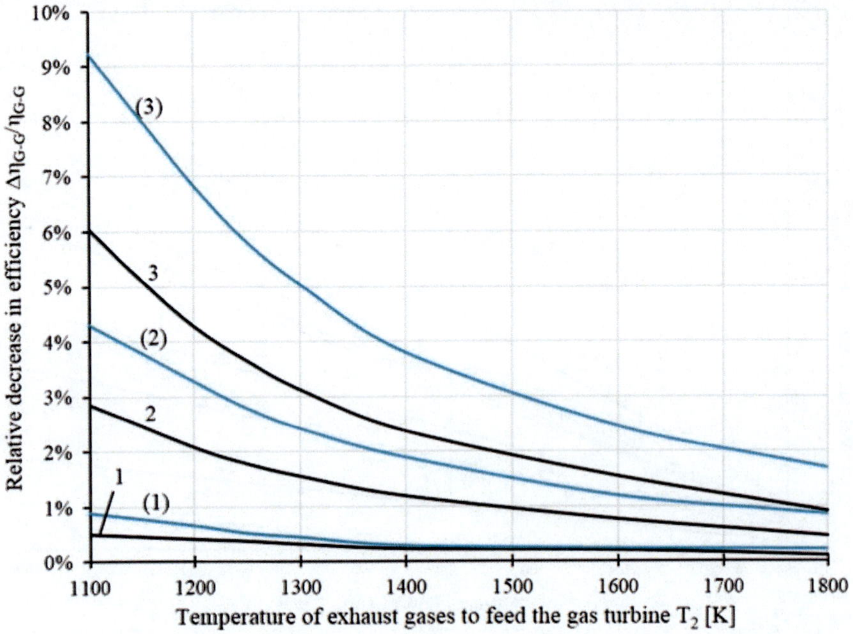

Fig. 6.8 Values of $\Delta\eta_{G-G}/\eta_{G-G}$ as a function of temperature T_2 for *Joule-Brayton* cycle with one- and two-stage compression with a pressure drop of Δp_{cav} as a parameter: 1—$\Delta p_{cav} = 0.1$ bar, 2—$\Delta p_{cav} = 0.5$ bar, 3—$\Delta p_{cav} = 1$ bar (values in parentheses apply to the *Joule-Brayton* cycle with two-stage compression)

and only the path from the general to specific is correct and gives such an opportunity. Thus, the innovative methodology [2–4] of recording profit in continuous time by means of the integral functional (6.22) gives a completely new quality and new possibilities of technical and economic analyses of all investment processes. "Continuous" notation (6.22) allowes:

- to obtain NPV functions for the processes under analysis, (6.35) and (6.36), whereas the discrete record gives only the detail (point), i.e. the numerical value of the NPV, which makes any analysis impossible. This can not be changed even by calculating a few dozen or more NPV values and considering the largest of them the optimal one.

The continuous methodology facilitates:

- using differential calculus to study the variability of the NPV function and thus to obtain comprehensive information about them, as well as to prepare a grapgh, which allowes obtaining a whole range of additional, important information that would be impossible withouth it, or at least it would be very difficult to see. Moreover, the differential calculus makes it easy to find extreme values and the maximum of the NPV function. The resulting functions therefore show the nature of the course of changes in NPV values.

Another extremely important issue. "Continuous" recording:

- makes it possible to use any sub-functions, i.e. any time scenarios (6.22) characterising the investment process under analysis, which allow, what is very important, to analyse the future, to create thinking about it in a scientific way.

To sum up, the functions NPV_{CAES} (6.35), (6.36) obtained by means of the functional (6.22) have a fundamental, even invaluable value. They give an overall picture of the process under analysis (Figs. 6.9 and 6.10), which could not be provided by a discrete record of the meter NPV_{CAES} and the "point" values obtained by means of it. These functions thus enable discussion and analysis of research results, and allow drawing conclusions of a general nature about the analysed investment process.

As can be seen from Fig. 6.3, the daily increase in revenue from the sale of electricity solely due to the use of a compressed air storage in the system is:

$$\Delta S = S_{CAES} - S_{no\,CAES} =$$

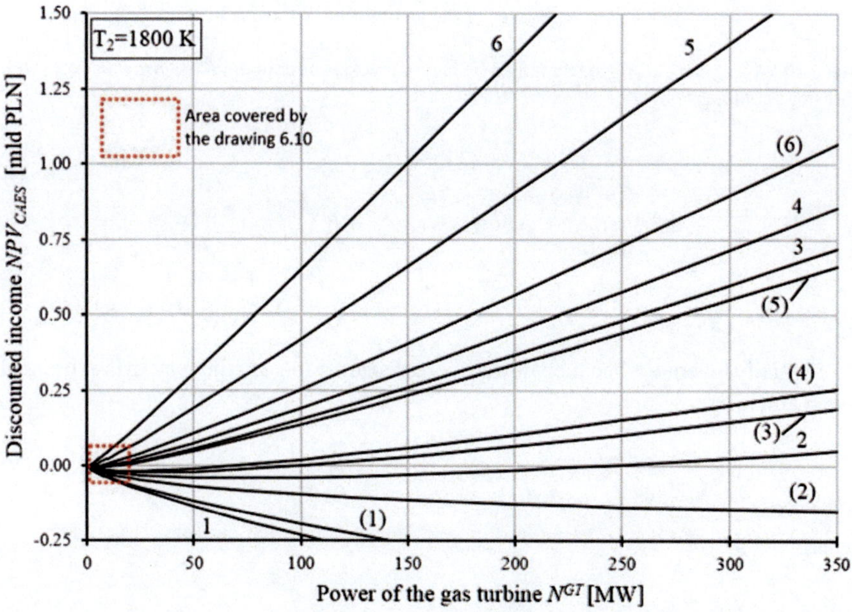

Fig. 6.9 Discounted profit NPV_{CAES} as a function of gas turbine power N^{GT} for *Joule-Brayton* cycles with one- and two-stage compression with the difference in peak and off-peak electricity prices $e^b_{el} - e^d_{el}$ and the off-peak duration τ_a as parameters, where: 1—$e^b_{el} - e^d_{el} = 30$ PLN, $\tau_a = 14$ h; 2—$e^b_{el} - e^d_{el} =$ PLN 50, $\tau_a = 14$ h; 3—$e^b_{el} - e^d_{el} = 70$ PLN, $\tau_a = 14$ h; 4—$e^b_{el} - e^d_{el} = 30$ PLN, $\tau_a = 12$ h; 5—$e^b_{el} - e^d_{el} = 50$ PLN, $\tau_a = 12$ h; 6—$e^b_{el} - e^d_{el} = 70$ PLN, $\tau_a = 12$ h (values in brackets refer to *Joule-Brayton* cycle with two-stage compression)

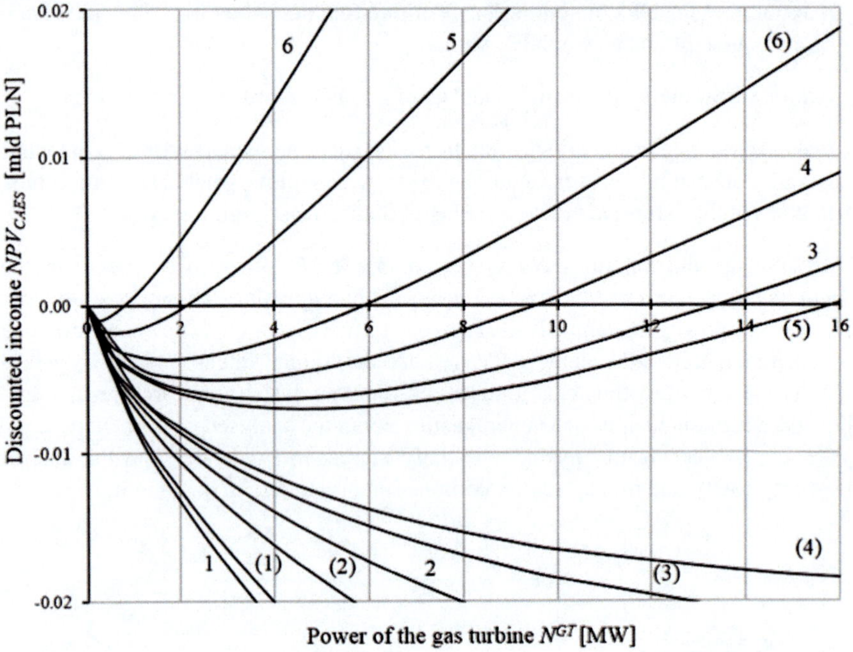

Fig. 6.10 A fragment of Fig. 6.9 enlarged (values in brackets refer to the *Joule-Brayton* cycle with two-stage compression)

$$= [(N^{G-G} - \Delta N_a)\tau_D e_{el}^d + N^b(\tau_D - \tau_a)e_{el}^b - N^{G-G}\tau_D e_{el}^d](1 - \varepsilon_{el}) =$$

$$= [\Delta N_a(\tau_D - \tau_a)(e_{el}^b - e_{el}^d) + \Delta N_b(\tau_D - \tau_a)e_{el}^b - \Delta N_a \tau_a e_{el}^d](1 - \varepsilon_{el}) \quad (6.18)$$

The daily revenue when there is a compressed air storage in the system is expressed by an equation:

$$S_{CAES} = \left[(N^{G-G} - \Delta N_a)\tau_D e_{el}^d + N^b(\tau_D - \tau_a)e_{el}^b\right](1 - \varepsilon_{el}) \quad (6.19)$$

where:

$$N^b = \Delta N_a + \Delta N_b \quad (6.20)$$

whereas the daily revenue when there is not a compressed air storage in the system is:

$$S_{no\ CAES} = N^{G-G}\tau_D e_{el}^d(1 - \varepsilon_{el}) \quad (6.21)$$

where:

e^b_{el}, e^d_{el} peak and off-peak electricity prices,
ε_{el} indicator of the electrical needs of the gas engine and cavern infrastructure (without, of course, the power of the compressor injecting air into the cavern).

At the peak of the demand for electricity, air from the cavern at a temperature T_{10} lower than the temperature T_2 is fed into the combustion chamber CH of the gas turbine. However, formulas (6.18)–(6.20) neglect influence of the difference of temperature on the power of a gas turbine N^{TG}, and thus on the power N^{G-G}.

The calculation assumes that the number of off-peak hours τ_a is equal to the number of peak hours: $\tau_a = \tau_b = \tau_D - \tau_a$. If $\tau_b > \tau_a$, the cost of purchasing electricity from the grid to cover the electrical needs during the time difference to power the motor driving the air compressor charging the cavern would have to be deducted from the revenue ΔS (formula (6.18)) to cover the total discounted profit from the use of a compressed air accumulator in the system over the years is:

$$NPV_{CAES} = \int_0^{T_y} \left[\Delta S_R - \Delta K_e - F^{CAES} - R^{CAES} - \left(\Delta S_R - \Delta K_e - F^{CAES} - A^{CAES}\right)p\right] \mathrm{e}^{-rt} dt \tag{6.22}$$

where:

A^{CAES} amortisation instalment,

$$A^{CAES} = J^{CAES} / T_y \tag{6.23}$$

J^{CAES} capital expenditure on adapting the cavern to work with a *gas-gas* engine
F^{CAES} interest (financial costs) on investment funds J^{CAES}, variable over time,

$$F^{CAES} = r[J^{CAES} - (t-1)R^{CAES}], \tag{6.24}$$

ΔK_e annual cost of maintenance and repair of cavern infrastructure,
p income tax rate,
R^{CAES} loan repayment installment,

$$R^{CAES} = J^{CAES} / T_y, \tag{6.25}$$

r stopa oprocentowania kapitału J^{aku},

ΔS_R time-varying growth of annual revenue from generating electricity at peak hours,

t time,

T_y the compressed air storage life expectancy expressed in years.

After taking into account all the indices, the annual revenue shall be:

$$\Delta S_R = \{L_z [\Delta N_a^z(\tau_D - \tau_a^z)(e_{el}^b - e_{el}^d) + \Delta N_b^z(\tau_D - \tau_a^z)e_{el}^b - \Delta N_a^z\tau_a^z e_{el}^d] +$$

$$+ L_l [\Delta N_a^l(\tau_D - \tau_a^l)(e_{el}^b - e_{el}^d) + \Delta N_b^l(\tau_D - \tau_a^l)e_{el}^b - \Delta N_a^l\tau_a^l e_{el}^d]\}(1 - \varepsilon_{el}), \tag{6.26}$$

where:

L_z number of winter days,
L_l number of other than winter days,
ΔN reduction/increase in the power of the gas/gas engine.

If additionally assume that the power increases in winter and outside winter are identical, $\Delta N^z = \Delta N^l$, and that the numbers of peak hours and off-peak hours are equal, $\tau_a^z = \tau_a^l$, is the increase in annual revenue from peak electricity generation is represented by the equation:

$$\Delta S_R = (365 - L)[\Delta N_a(\tau_D - \tau_a)(e_{el}^b - e_{el}^d) + \Delta N_{sz}(\tau_D - \tau_a)e_{el}^b - \Delta N_a\tau_a e_{el}^d](1 - \varepsilon_{el}) \tag{6.27}$$

where:

L number of days per year when the engine is not running.

The first component on the right hand side of the above dependence expresses an increase in revenue from the sale of peak electricity resulting solely from the difference between peak and off-peak electricity prices (in the system without a storage the electricity $\Delta N_a(\tau_D - \tau_a)$ is sold at the price e_{el}^d), the second component is the revenue from the sale of electricity at a peak price, the third component is the opportunity cost of not selling electricity in off-peak hours.

If, moreover, it is assumed that the increase in power ΔN equals the compression power $N_{i\ C}$, i.e. that the air from the cavern is fed directly to the combustion chamber of the turbine during peak hours, and that during off-peak hours the power $N_{i\ C}$ is used to drive the compressor that charges the cavern, then:

for *Joule-Brayton* cycle with single-stage compression

$$\Delta S_R = (365 - L)[N^{GT}\frac{z_{GT}^{opt}T_{amb}}{\eta_m^2\eta_i^C\eta_i^{GT}T_2 - z_{GT}^{opt}T_{amb}}(\tau_D - \tau_a)(e_{el}^b - e_{el}^d) +$$

$$+N^{GT}\frac{z_{GT}^{opt}T_{amb}}{\eta_m^2\eta_i^C\eta_i^{GT}T_2-z_{GT}^{opt}T_{amb}}(\tau_D-\tau_a)e_{el}^b-N^{GT}\frac{z_{GT}^{opt}T_{amb}}{\eta_m^2\eta_i^C\eta_i^{GT}T_2-z_{GT}^{opt}T_{amb}}\tau_a e_{el}^d](1-\varepsilon_{el}) \tag{6.28}$$

for *Joule-Brayton* cycle with two-stage compression

$$\begin{aligned}\Delta S_R=(365-L)[N^{GT}&\frac{(z_{GT}^{opt})^2(T_{amb}+T_4)}{\eta_m^2\eta_i^C\eta_i^{GT}T_2(z_{GT}^{opt}+1)-(z_{GT}^{opt})^2(T_{amb}+T_4)}(\tau_D-\tau_a)(e_{el}^b-e_{el}^d)+\\ +N^{GT}&\frac{(z_{GT}^{opt})^2(T_{amb}+T_4)}{\eta_m^2\eta_i^C\eta_i^{GT}T_2(z_{GT}^{opt}+1)-(z_{GT}^{opt})^2(T_{amb}+T_4)}(\tau_D-\tau_a)e_{el}^b+\\ -N^{GT}&\frac{(z_{GT}^{opt})^2(T_{amb}+T_4)}{\eta_m^2\eta_i^C\eta_i^{GT}T_2(z_{GT}^{opt}+1)-(z_{GT}^{opt})^2(T_{amb}+T_4)}\tau_a e_{el}^d](1-\varepsilon_{el})\end{aligned} \tag{6.29}$$

The investment expenditure on the entire cavern infrastructure J^{CAES} can be expressed, for example, by means of a formula:

$$J^{CAES}=N^{G-G}i^{CAES} \tag{6.30}$$

and assuming that $N^{G-G}\approx 1,15N^{\mathrm{GT}}$(see Fig. 3.11) and that the unit investment expenditure (per unit of gas engine power) for the cavern infrastructure i^{CAES} is equal to the unit investment expenditure for the gas turbine set $i^{CAES}=i^{GT}$:

$$i^{GT}=1579(N^{\mathrm{GT}})^{-0,341}[\mathrm{USD/kW}] \tag{6.31}$$

(the capital expeditures were determined by means of [6] with the power N^{GT} expressed in MW), we obtain:

$$J^{CAES}=\mathrm{A}(N^{\mathrm{GT}})^{\mathrm{B}}\ [\mathrm{mln\ USD}] \tag{6.32}$$

where:

A = 1,8
B = 0,659.

Assuming the annual cost of maintenance and repair of the cavern infrastructure can be expressed using the standard dependency:

$$\Delta K_e=\delta_{rem}J^{CAES} \tag{6.33}$$

(the calculations are based on $\delta_{rem}=0{,}03$),

and assuming electricity prices $e_{el}(t)$ for example by means of an exponential function:

$$e_{el}(t) = e_{el}^{t=0} e^{a_{el} t} \tag{6.34}$$

(depending on the value a_{el} the price $e_{el}(t)$ may increase, decrease or be a constant value in subsequent years) and by substituting all the quantities (formulae (6.23)–(6.25), (6.28)–(6.34)) in the formula (6.22) and then performing the integration, the final forms of dependence on discounted profit NPV_{CAES} are earned over the years T_y from the storage operation as a function of gas turbine power N^{GT}:

- for *Joule-Brayton* cycle with single-stage compression

$$NPV_{CAES} = \left\{ N^{GT} \frac{z_{GT}^{opt} T_{amb}(\tau_D - \tau_a)}{\eta_m^2 \eta_i^C \eta_i^{GT} T_2 - z_{GT}^{opt} T_{amb}} \left[e_{el}^{b,t=0} \frac{e^{(a_{el}^b - r)T_y} - 1}{a_{el}^b - r} - e_{el}^{d,t=0} \frac{e^{(a_{el}^d - r)T_y} - 1}{a_{el}^d - r} \right] + \right.$$
$$\left. + N^{GT} \frac{z_{GT}^{opt} T_{amb}}{\eta_m^2 \eta_i^C \eta_i^{GT} T_2 - z_{GT}^{opt} T_{amb}} \left[(\tau_D - \tau_a) e_{el}^{b,t=0} \frac{e^{(a_{el}^b - r)T_y} - 1}{a_{el}^b - r} - \tau_a e_{el}^{d,t=0} \frac{e^{(a_{el}^d - r)T_y} - 1}{a_{el}^d - r} \right] \right\} \times$$
$$\times (365 - L)(1 - \varepsilon_{el})(1 - p) - \mathrm{A}(N^{GT})^{\mathrm{B}} \left[(1 - e^{-rT_y}) \frac{\delta_{rem}}{r} + z_z \left(\frac{1 - e^{-rT_y}}{T_y} + 1 \right) \right] (1 - p) \tag{6.35}$$

- for *Joule-Brayton* cycle with two-stage compression

$$NPV_{CAES} =$$
$$\left\{ N^{GT} \frac{(z_{GT}^{opt})^2 (T_{amb} + T_4)(\tau_D - \tau_a)}{\eta_m^2 \eta_i^C \eta_i^{GT} T_2 (z_{GT}^{opt} + 1) - (z_{GT}^{opt})^2 (T_{amb} + T_4)} \left[e_{el}^{b,t=0} \frac{e^{(a_{el}^b - r)T_y} - 1}{a_{el}^b - r} - e_{el}^{d,t=0} \frac{e^{(a_{el}^d - r)T_y} - 1}{a_{el}^d - r} \right] + \right.$$
$$\left. + N^{GT} \frac{(z_{GT}^{opt})^2 (T_{amb} + T_4)}{\eta_m^2 \eta_i^C \eta_i^{GT} T_2 (z_{GT}^{opt} + 1) - (z_{GT}^{opt})^2 (T_{amb} + T_4)} \left[(\tau_D - \tau_a) e_{el}^{b,t=0} \frac{e^{(a_{el}^b - r)T_y} - 1}{a_{el}^b - r} - \tau_a e_{el}^{d,t=0} \frac{e^{(a_{el}^d - r)T_y} - 1}{a_{el}^d - r} \right] \right\} \times$$
$$\times (365 - L)(1 - \varepsilon_{el})(1 - p) - \mathrm{A}(N^{GT})^{\mathrm{B}} \left[(1 - e^{-rT_y}) \frac{\delta_{rem}}{r} + z_z \left(\frac{1 - e^{-rT_y}}{T_y} + 1 \right) \right] (1 - p) \tag{6.36}$$

with $e_{el}^b - e_{el}^d$, T_2 as parameters (for a *Joule-Brayton* cycle with two-stage compression, temperature T_4 is an additional parameter). The calculation assumes that $a_{el} = 0$, which should of course be interpreted as the electricity price is the overall average price over the years T_y and equals:

$$e_{el}^{mean} = \frac{1}{T_y} \int_0^{T_y} e_{el}^{t=0} e^{a_{el} t} dt = \frac{e_{el}^{t=0}}{T_y a_{el}} (e^{a_{el} T_y} - 1) = e_{el}^{t=0}. \tag{6.37}$$

The z_z in formulae (6.35), (6.36) is the capital J^{aku} freeze (capital J^{aku} during the construction of the battery is not profitable; $z_z > 1$) [2–5, 7–11].

As already mentioned above, the "continuous" time record of the function $NPV_{CAES} = f(N^{GT})$ (formula (6.22) and the resulting from it formulas (6.35),

(6.36)) enables the analysis of its course by means of a differential calculus. For example, the second derivative d^2NPV_{CAES}/dN^{GT^2} takes only positive values. The function $NPV_{CAES} = f(N^{GT})$ is therefore a concave function throughout the entire power variation range of a gas turbine $N^{GT} \in \langle 0; \infty)$ and its minimum value always assumes a negative value: $NPV_{CAES}^{\min} = f(N_{\min}^{GT}) < 0$. The function $NPV_{CAES} = f(N^{GT})$ is therefore strictly decreasing in the range $N^{GT} \in \langle 0; N_{\min}^{GT})$ and constantly increasing in the range $N^{GT} \in (N_{\min}^{GT}; \infty)$. The value of NPV_{CAES} goes to infinity when the power N^{GT} goes to infinity: $NPV_{CAES} \to \infty$ when $N^{GT} \to \infty$. The value of N_{min}^{GT} can of course be calculated from the necessary condition for existence of an extremum of a differentiable function $dNPV_{CAES}/dN^{GT} = 0$. Examples of curves $NPV_{CAES} = f(N^{GT})$ for the actual power range of the turbines produced $N^{GT} \in \langle 0; 350)$ is shown in Figs. 6.9 and 6.10. From the observations of some of them, one could draw an erroneous conclusion, resulting from the aforementioned limited power range N^{GT} in practice that they are strictly decreasing. This is only true for this limited range, since in this case $NPV_{CAES} \to \infty$ when $N^{GT} \to \infty$. Negative values of NPV_{CAES} for these curves (e.g. curve (2) in Fig. 6.9) show, however, that for the entire range $N^{GT} \in \langle 0; 350)$ of produced turbines accumulation of electricity for given values $e_{el}^b - e_{el}^d$, τ_a, J^{CAES} may be economically unprofitable.

6.3.1 Discounted Profit from the Use of a Compressed Air Storage

Figure 6.9 shows the results of multi-variant calculations using formulae (6.35) and (6.36) of the value of the profit NPV_{CAES} achieved from the operation of a compressed air storage cooperating with a hierarchical *gas-gas* engine. In the calculations, the same capital expeditures were assumed for both engine variants (formula (6.32)).

As shown in Fig. 6.9, the use of the engine in cooperation with a compressed air storage for the value of $e_{el}^b - e_{el}^d = 30$ PLN and time $\tau_a = 14$ h is economically unprofitable. Namely, profit NPV_{CAES} is always negative in the entire power range $N^{GT} \in \langle 0; 350)$ of gas turbines produced. For $e_{el}^b - e_{el}^d = 50$ PLN and $\tau_a = 14$ h, the use of the storage is profitable only above the rated power $N_{TG} = 250$ MW (see curve 2 in Fig. 6.9), and in the case of $e_{el}^b - e_{el}^d =$ PLN 30 and $\tau_a = 12$ h for example the storage is profitable only above the rated power $N_{TG} = 9.5$ MW (see curve 4 in Fig. 6.10). Increasing the value of $e_{el}^b - e_{el}^d$ and especially shortening the time τ_a increases this profitability. For the off-peak duration $\tau_a <= 12$ h and for the difference in peak and off-peak electricity prices $e_{el}^b - e_{el}^d > 30$ PLN it is always profitable to use a battery, even in the case of a gas turbine of a rated power N_{TG} in the order of several dozen kilowatts. Moreover, the use of a gas engine with two-stage compression and intercooling is less cost-effective than using an engine with single-stage compression. It results from the small values for the "two-stage" engine of the ratios $N_{i\,C}/N_{i\,exp}$ i $N_{i\,C}/N_{GT}$ (formulas (6.11), (6.12))—Fig. 6.4.

For example, for a gas turbine rated power $N_{TG} = 200$ MW and a temperature of $T_2 = 1800$ K, the value of the ratio $N_{i\ C}/N_{i\ exp}$ is 0.43 in a two-stage compression system, while in a one-stage compression system it amounts to as much as 0.60, so the compressor "eats" as much as 60% of the power $N_{i\ exp}$ of expansion in the turbine, which significantly reduces its power N^{GT} ($N^{GT} = N_{i\ exp} - N_{i\ C}$). So, despite the fact that the "two-stage" engine is thermodynamically more favorable, the peak power $N^b = 2N_{i\ C}$—Fig. 6.3—is also small, which consequently translates into a lower income from the sale of peak electricity (formula (6.29)), and thus a lower profit NPV_{CAES} (formula (6.36)).

6.3.1.1 Sensitivity Analysis of Discounted Profit

The analysis of the economic profitability of a business undertaking should be supplemented with a sensitivity analysis of its characteristic measures in order to assess changes in their value as a function of changes in the parameters affecting them. The future is unpredictable and the so-called baseline values (e.g. capital expenditure, electricity price levels, etc.) may be different in the future. Therefore, the sensitivity analysis gives the investor a large field of view of the investment profitability. Moreover, it allows to assess its "safety" were the values adopted for the economic analysis of the input data going to change. In a competitive environment, it also enables pricing policy.

Figure 6.11 shows the results of a sensitivity analysis of the discounted profit NPV_{aku} to the change in the difference between peak and off-peak electricity prices, the duration of off-peak, the capital expeditures on the compressed air storage, interest rates and the power of the gas turbine. The values of these quantities were changed in the range of ± 20% from their base values. The base values were: $e^b_{el} - e^d_{el} = 50$ PLN/MWh, $\tau_a = 12$ h, $J^{CAES} = 214$ mlnPLN, r = 6%, $N^{GT} = 202$ MW(Siemens SGT6-5000F turbine). The reduced values in Fig. 6.12 corresponding to the base values take the value of 1 on the abscissa, of course, and on the ordinate the profit values for the gas turbine with one- and two-stage compression, $NPV_{CAES} = 0,91$ mld PLN and $NPV_{CAES} = 0,34$ mld PLN respectively.

As shown in Fig. 6.11, the greatest impact on the value of profit NPV_{CAES} has the daily duration of off-peak time τ_a. The shorter it is, i.e. the longer the peak time τ_b, the higher the revenue from the sale of the peak electricity. The time τ_a is therefore fundamental to the profitability of electricity storage. After all, the idea of sstoring electricity over time τ_a is the basis for the use of compressed air accumulators. The second most important factor influencing NPV_{CAES} is the difference peak and off-peak electricity prices $e^b_{el} - e^d_{el}$. The greater it is, the greater the profit of NPV_{CAES} is of course. Increasing the power of the *gas-gas* engine, i.e. increasing the power of the gas turbine N^{GT} also increases the profit (formulas (6.35), (6.36)), as the sales of peak electricity are higher then. On the other hand, the profit of NPV_{CAES} is influenced to a small extent by investment capital J^{CAES}. This is due to the relatively small annual capital cost (the financial cost plus depreciation) $K_{cap} = F^{CAES} + A^{CAES}$ of storage operation, which depends on these inputs. Depending on the rate r and

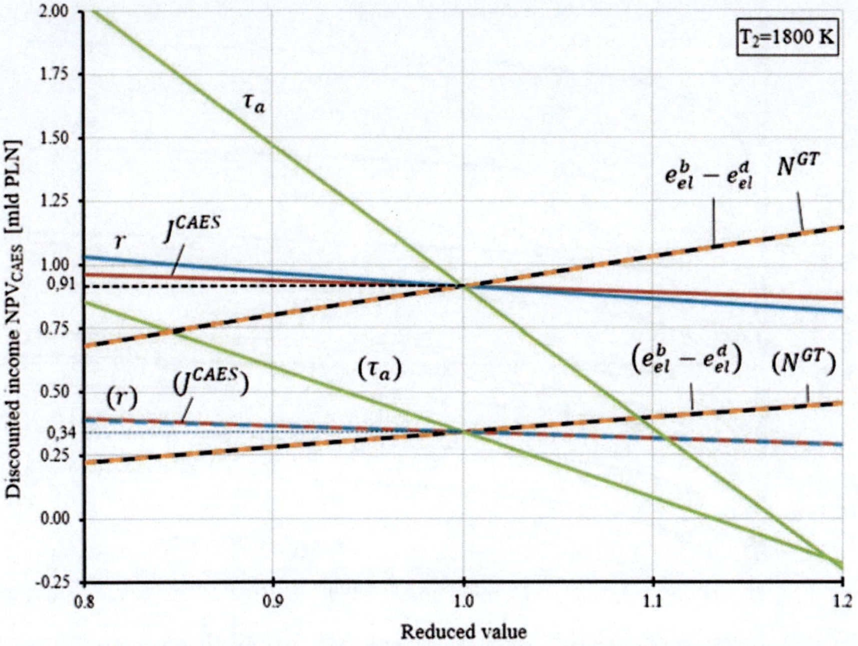

Fig. 6.11 The impact of the difference in peak and off-peak electricity prices $e^b_{el} - e^d_{el}$, the duration of off-peak τ_a, the value of investment capital J^{CAES}, the rate r of its interest and the power of the gas turbine N^{GT} on the discounted profit NPV_{CAES} achieved from the operation of the compressed air storage cooperating with *gas-gas* engine with *Joule-Brayton* cycles of a gas turbine with one and two-stage compression (values in brackets refer to the *Joule-Brayton* cycle with two-stage compression)

T_y years of the storage life, the annual cost K_{cap} is only a few (at most a dozen) percent of the expenditure J^{CAES} (of course increasing the rate r increases this cost, but it decreases with the increase in T_y years). Total capital costs over T_y years are presented in Eqs. (6.35) and (6.36) by $K_{kap} = J_{CAES}\, z_z\, \{[1 - \exp(-rT_y)]/T_y + 1\}$.

The results of the calculations of the profit value NPV_{CAES} presented in Fig. 6.9 were performed for L = 30 days a year, when the engine does not operate. Even doubling the value of L negligibly reduces the value of NPV_{CAES}. For this reason, Fig. 6.11 does not show the line of changes in the value of NPV_{CAES} with a change in the value of L. They would overlap with the "base" lines $NPV_{CAES} =$ PLN 0.91 and PLN 0.34 billion. For the same reason, Fig. 6.14 does not show changes in the value of $\Delta k_{el,\,mean}$ (formula (6.39)) with a change in the value of L (the "baselines" in this case: $\Delta k_{el,\,mean} = 53.14$ and 18.59 PLN/MWh).

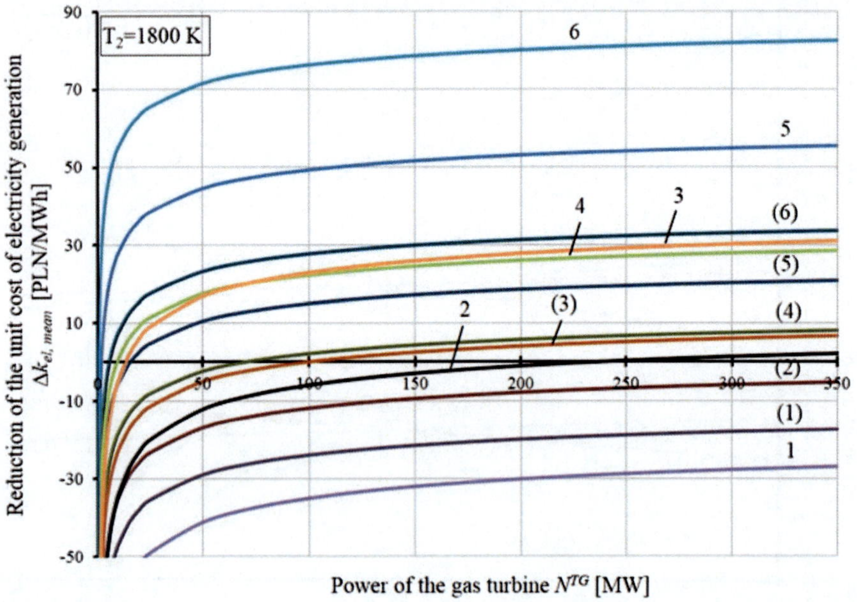

Fig. 6.12 Reduction of electricity generation unit cost $\Delta k_{el,\,mean}$ as a function of gas turbine power N^{GT} for temperature $T_2 = 1800$ K for *Joule-Brayton* cycle with single-stage and dual-stage compression with the difference in peak and off-peak electricity prices $e^b_{el} - e^d_{el}$ and off-peak duration τ_a as parameters where: 1—$e^b_{el} - e^d_{el} = 30$ PLN, $\tau_a = 14$ h; 2—$e^b_{el} - e^d_{el} = 50$ PLN, $\tau_a = 14$ h; 3—$e^b_{el} - e^d_{el} = 70$ PLN, $\tau_a = 14$ h; 4—$e^b_{el} - e^d_{el} = 30$ PLN, $\tau_a = 12$ h; 5—$e^b_{el} - e^d_{el} = 50$ PLN, $\tau_a = 12$ h; 6—$e^b_{el} - e^d_{el} = 70$ PLN, $\tau_a = 12$ h (values in brackets refer to the *Joule-Brayton* cycle with two-stage compression)

6.3.2 The Influence of the Use of Compressed Air Storage in a System with a Gas-Gas Engine on the Reduction of the Unit Cost of Electricity Generation

Integration the discounted profit NPV_{CAES} (formulas (6.35), (6.36)) expressed by the equation:

$$NPV_{CAES} = \int_0^{T_y} E_{el,R}\Delta k_{el,\,mean}(1-p)\mathrm{e}^{-rt}dt = E_{el,R}\Delta k_{el,\,mean}(1-p)\int_0^{T_y} \mathrm{e}^{-rt}dt, \tag{6.38}$$

yields a formula for the average reduction in T_y years of the unit cost of electricity generation in a gas engine owing to its cooperation with a compressed air storage by value:

$$\Delta k_{el,\,mean} = \frac{NPV_{CAES}}{(1-p)E_{el,R}} \frac{r}{1-\mathrm{e}^{-rT_s}} \tag{6.39}$$

with the annual electricity generation $E_{el,R}$ by the system being expressed in the formula (cf. formulae (6.19), (6.27)):

$$E_{el,R} = (365 - L)[(N^{G-G} - \Delta N_a)\tau_D + N^b(\tau_D - \tau_a)](1 - \varepsilon_{el}) \tag{6.40}$$

and as assumed above:

$$N^b = \Delta N_a + \Delta N_b = N_{i\,C} + N_{i\,C} = 2N_{i\,C} \tag{6.41}$$

$$N^{G-G} = N^{GT} + N^{TE} = N^{GT}\left(1 + \frac{N^{TE}}{N^{GT}}\right) \tag{6.42}$$

so that:

$$E_{el,R} = (365 - L)\left\{\left[N^{GT}\left(1 + \frac{N^{TE}}{N^{GT}}\right) - N_{i\,C}\right]\tau_D + 2N_{i\,C}(\tau_D - \tau_a)\right\}(1 - \varepsilon_{el}). \tag{6.43}$$

The ratio value N^{TE}/N^{GT} is shown in Fig. 6.4.

Using the formula (6.39), a multi-variant calculation of the unit cost reduction $\Delta k_{el,\,mean}$ was performed, the results of which are presented in Figs. 6.12 and 6.13. Figure 6.12 shows the results for the temperature of the exhaust gas supplied from the combustion chamber to the gas turbine equal to $T_2 = 1800$ K, Fig. 6.13 for $T_2 = 1400$ K.

The values of $\Delta k_{el,\,mean}$ for off-peak duration $\tau_a = 12$ h are very significant. The values shown in Fig. 6.12 $\Delta k_{el,\,mean}$ were obtained for the highest temperature currently feasible due to the limited heat resistance of the turbine blades $T_2 = 1800$ K, in Fig. 6.13 for $T_2 = 1400$ K. The values of $\Delta k_{el,\,mean}$ increase with decreasing temperature values T_2, which is obviously economically advantageous. This is because the smaller the value T_2, the higher the compression power $N_{i\,C}$, Fig. 6.4, (which in turn, of course, is thermodynamically unfavourable), so there is more peak power N^b (formula 6.41) and thus the profit is higher NPV_{CAES}.

The values of $\Delta k_{el,\,mean}$ will be even higher τ_a for time τ_a less than 12 h and a difference of peak and off-peak electricity prices $e_{el}^b - e_{el}^d$ greater than 30 PLN. The unit cost of electricity generation in a *gas-gas* engine cooperating with a compressed air storage (Fig. 6.1) will therefore, what is extremely important, very significantly lower than the unit cost of $k_{el,\,mean}^{G-P}$ in a gas-steam engine (Sect. 3.3.2; Figs. 3.22 and 3.23). This cost in a *gas-gas* engine cooperating with a compressed air storage will be:

$$k_{el,\,mean,\,CAES}^{G-G} = k_{el,\,mean}^{G-G} - \Delta k_{el,\,mean} < k_{el,\,mean}^{G-P} \tag{6.44}$$

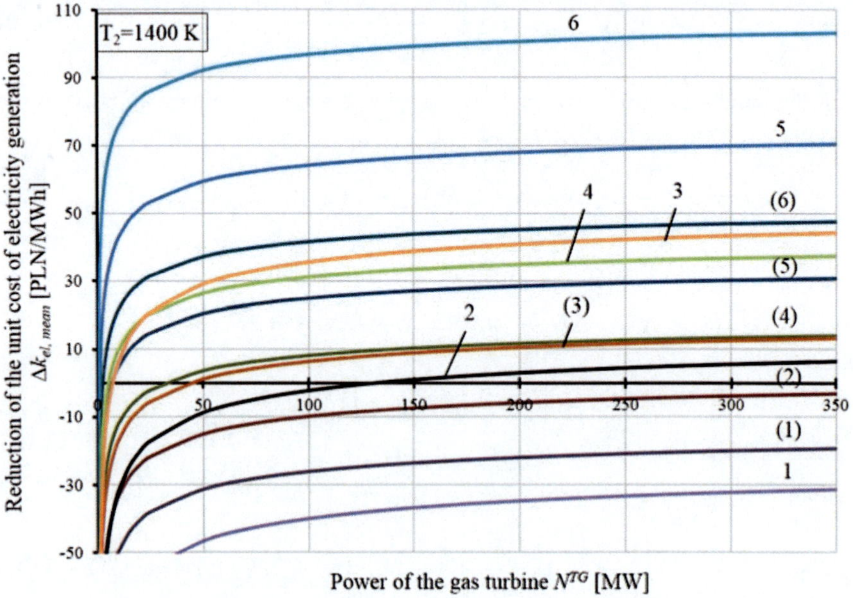

Fig. 6.13 Reduction of the unit cost of electricity generation $\Delta k_{el,mean}$ as a function of gas turbine power N^{GT} for temperature $T_2 = 1400$ K for *Joule-Brayton* cycle with single-stage and dual-stage compression with the difference in peak and off-peak electricity prices $e_{el}^{sz} - e_{el}^{d}$ and off-peak duration τ_a as parameters where: 1—$e_{el}^{b} - e_{el}^{d} = 30$ PLN, $\tau_a = 14$ h; 2—$e_{el}^{b} - e_{el}^{d} = 50$ PLN, $\tau_a = 14$ h; 3—$e_{el}^{sz} - e_{el}^{d} = 70$ PLN, $\tau_a = 14$ h; 4—$e_{el}^{b} - e_{el}^{d} = 30$ PLN, $\tau_a = 12$ h; 5—$e_{el}^{b} - e_{el}^{d} = 50$ PLN, $\tau_a = 12$ h; 6—$e_{el}^{b} - e_{el}^{d} = 70$ PLN, $\tau_a = 12$ h (values in brackets refer to the *Joule-Brayton* cycle with two-stage compression)

and it will be the lower than $k_{el,\,mean}^{G-P}$ the lower the temperature T_2. Also, the unit cost of heat production in a gas-and-gas CHP plant (see Sect. 3.3.1; Figs. 3.13, 3.14, 3.15, 3.16, 3.17, 3.18, 3.19, 3.20 and 3.21) will be even more lower than the cost of a *gas-steam* CHP plant. The revenue from the sale of electricity will increase significantly, which is the avoided cost of heat production.

The values $\Delta k_{el,mean}$, as well as the discounted profit values NPV_{CAES} for the gas turbine engine in which the *Joule-Brayton* cycle is performed with two-stage compression, are lower compared to the single stage compression.

6.3.2.1 Sensitivity analysis of the reduction in the unit cost of electricity generation

Figure 6.14 presents an analysis of the sensitivity of the reduction in the unit cost of electricity generation $\Delta k_{el,mean}$ to the change in the difference between peak and off-peak electricity prices, duration of off-peak, the capital expeditures on the compressed air storage, the interest rate and the power of the gas turbine. The base values adopted

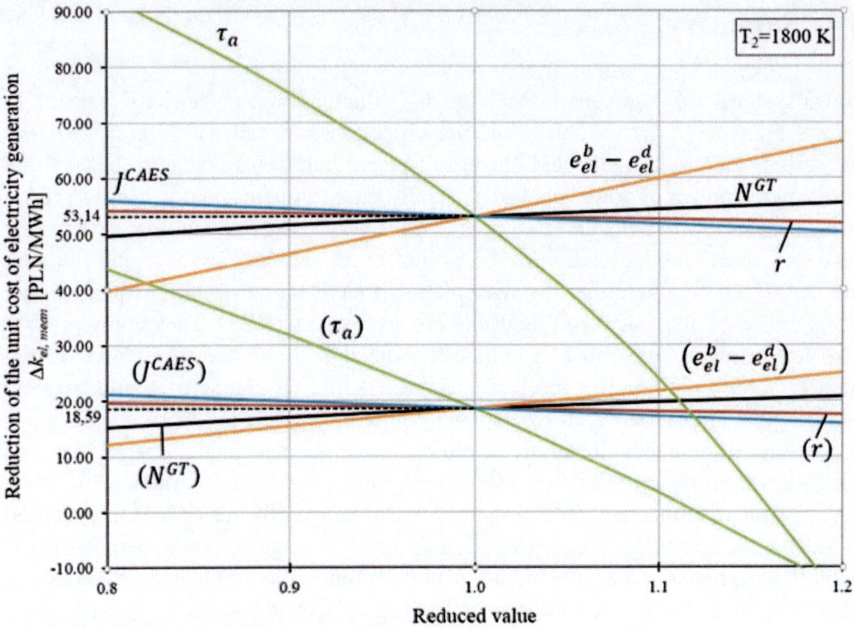

Fig. 6.14 The impact of the difference in peak and off-peak electricity prices $e_{el}^b - e_{el}^d$, the off-peak duration τ_a, the value of investment capital J^{CAES}, the r rate of interest and the power of the gas turbine N^{GT} on the reduction of the unit cost of electricity generation in a *gas-gas* engine with *Joule-Brayton* cycles of a gas turbine with single and two-stage compression (values in parentheses refer to the *Joule-Brayton* cycle with two-stage compression)

here are identical to those in the discounted profit NPV_{CAES} sensitivity analysis (Sect. 6.3.1.1).

As shown in Fig. 6.14, the off-peak duration τ_a has the greatest impact on the value of $\Delta k_{el,mean}$ (as well as on the value of NPV_{CAES} shown in Fig. 6.11). The shorter it is, i.e. the longer the peak time of increased needs, the greater the reduction of the unit cost. The second largest factor influencing is the difference in peak and off-peak electricity prices $e_{el}^b - e_{el}^d$. The greater it is, of course, the greater the cost reduction. On the other hand, the investment outlays for the compressed air storage infrastructure have little impact (for the same reasons as described in the profit sensitivity analysis—Sect. 6.3.1.1).

6.4 Summary

Power systems are characterized by large daily fluctuations of electricity demand. In order for power plants to be able to operate with a constant load, it is necessary to store the surplus electricity generated during off-peak hours. This is due to the fact that reducing the power of power plants, especially their frequent shutdown and start-up, which lasts a few to a dozen-or-so hours and is very energy-consuming at the same time, is a bad solution, leading to their rapid technical wear and tear and frequent failures. Unstable operation of power plants, both at the peak and off-peak hours, is magnified by the so-called renewable energy sources (RES). They enjoy priority dispatch in electricity grid and, which is especially important, are characterized by a very high randomness in electricity generation. Of course, it would be most advantageous not to store electricity and adjust the generation from power plants depending to the needs, if only it would not cause their technical degradation and a significant efficiency reduction when operating at part-laod. In practice, however, degradation and efficiency reduction are a reality. Currently, the only viable method of storing electricity for a long time during hours of reduced power demand is the potential energy of water stored in the upper pump storage tanks of hydropower plants. Their construction, however, is limited by terrain conditions and therefore such power plants are few and far between. Other ways of accumulating electricity have not been technically mastered so far.

One, however, apart from pumped-storage power plants, a rational method of storing electricity in the event of unstable operation of its sources caused by daily fluctuations in demand for power, further deepened by the unpredictable operation of renewable energy sources, may be its accumulation with compressed air in natural voids in the rock mass (caverns) or in underground mining workings.The analyses presented in this chapter show that this method of storage is justified both thermodynamically and economically. The use of gas turbine sets in which *Joule-Brayton* cycle with single-stage compression and expansion is carried out is highly justified (Fig. 4.3). Due to constructional simplicity such turbosets are produced. Thermodynamically improved engines, such as the engine in which the *Joule-Brayton* cycle with two-stage compression and intercooling is carried out, Fig. 6.2, are less economically viable (Figs. 6.9, 6.10, 6.11, 6.12, 6.13 and 6.14). *Joule-Brayton* cycle (Fig. 6.15), that is closer to the ideal, would be even less economically feasible.

This is because the peak power, Fig. 6.3, is then lower and lower, which translates into lower revenues from peak electricity sales. The most ideal, theoretical Ericson cycle, would be unprofitable to the highest degree (zero income from electricity sales, Ericson cycle is a special case of the generalised *Carnot* cycle). This is the isothermal compression and expansion cycle, Fig. 6.15, and therefore it refers to infinite number of compression and expansion stages. Thus, the least thermodynamically perfect *Joule-Brayton* cycle, i.e. a cycle with only one stage compression and expansion, turns out to be the most favorable economically. Paradoxically, therefore, gas engines that realize low efficiency cycles, i.e. characterised by a low inlet gas turbine flue gas temperature T_2, are economically the most advantageous, and the lower the

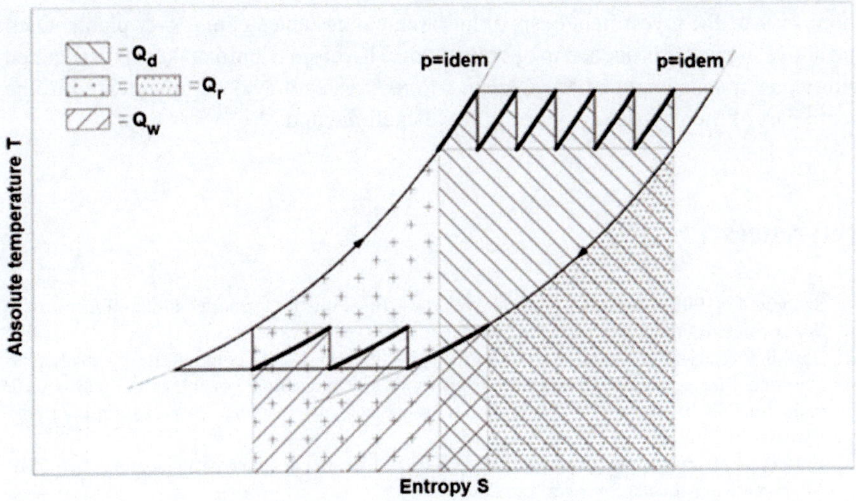

Fig. 6.15 Gas turbine cycle with heat regeneration, three-stage compression and six-stage expansion (Q_d- heat added to the cycle; Q_r- regeneration heat; Q_w- heat removed from the cycle)

temperature T_2 the more economically advantageous they are (see Figs. 6.12 and 6.13). Therefore, the proverb quoted in Chap. 5 is confirmed that *every cloud has a silver lining*. A thesis can be formulated, which is justified by the results of technical and economic calculations presented in the chapter, that: economics "stands up for inferior" technical solutions.

It is necessary to say once again, expressis verbis, that it would be most advantageous to give up electricity storage and use the most thermodynamically perfect engines for its generation, i.e., with the highest energy efficiency. Even if they will operate at part-load during off-peak hours, their reduced efficiency will still be higher than the efficiency of less perfect engines. This would translate into lower fuel consumption. Moreover, then, what is important, significant capital expenditure on electricity storage facilities would also be unnecessary, which would more than compensate for the higher expenditure on better engines. Equally important, numerous technical, ecological, landscape and social problems resulting from the construction of warehouse infrastructure would also be eliminated. Although they are difficult to estimate economically, their importance in investment processes is equally important and cannot be overestimated.In addition to stabilisation of the engine operation, the additional benefit of storing electricity with compressed air is a very significant reduction in the unit electricity generation cost in a hierarchical system of *gas-gas* engine + storage compared to stand-alone operation of the engine. This cost, which is extremely important, is also lower, and very significantly, than the cost of electricity generation in a hierarchical gas-steam engine. Therefore, where there are underground reservioirs allowing for the storage of compressed air, it is necessary to use inexpensive hierarchical *gas-gas* engine. This is due to the fact that its full construction/investment costs per unit of the plant capacity represent only about 45% of the investment costs of hierarchical gas and steam power plants. Over and above, no water is needed to operate them. Therefore, significant costs associated with water management of the *Clausius-Rankine* steam cycle, implemented in the steam part of the gas-steam systems, are also eliminated.

References

1. Szargut J (1998) Technical thermodynamics (in Polish: Termodynamika Techniczna). Wydawnictwo Politechniki Śląskiej, Gliwice
2. Bartnik R, Buryn Z, Hnydiuk-Stefan A (2017) Methodology and a continuous time mathematical model for selecting the optimum capacity of a heat accumulator integrated with a CHP plant (in Polish: Ekonomika Energetyki w modelach matematycznych z czasem ciągłym). Wydawnictwo Naukowe PWN, Warszawa
3. Bartnik R, Buryn Z, Hnydiuk-Stefan A (2017) Investment strategy in heating and CHP. Mathematical Models, Springer, London
4. Bartnik R, Bartnik B, Hnydiuk-Stefan A (2016) Optimum investment strategy in the power industry. Mathematical Models, Springer, New York
5. Bartnik R, Bartnik B (2014) Economic calculation in the energy sector (in Polish: Rachunek ekonomiczny w energetyce). WNT, Warszawa
6. Gas Turbine World (2007–2008) GTW Handbook. Volume 26, Pequot Publication, Inc. Southport, USA
7. Bartnik R (2009) Gas-steam power plants and combined heat and power plants. Energy and economic efficiency (in Polish: Elektrownie i elektrociepłownie gazowoparowe. Efektywność energetyczna i ekonomiczna), WNT, Warszawa (reprint 2012, 2017)
8. Bartnik R, Buryn Z, Hnydiuk-Stefan A, Skomudek W (2020) Power plant retrofit and modernization (in Polish: Modernizacja elektrowni. Efektywność energetyczna i ekonomiczna), WNT, Warszawa
9. Bartnik R, Buryn Z, Hnydiuk-Stefan A, Skomudek W (2019) Dual-Fuel Gas-Steam Combined Heat and Power Plants (in Polish: Dwupaliwowe elektrownie i elektrociepłownie gazowoparowe. Efektywność energetyczna i ekonomiczna), WNT, Warszawa
10. Bartnik R (2013) The modernization potential of gas turbines in the coal-fired power industry. thermal and economic effectiveness, Springer, London
11. Bartnik R, Buryn Z (2011) Conversion of coal-fired power plants to cogeneration and combined-cycle: thermal and economic effectiveness. Springer, London

Chapter 7
Replacing Natural Gas in a Gas–Gas Engine with Nuclear Fuel

Building nuclear power is absolutely necessary for a number of reasons. (1) Nuclear power is a carbon-free source of electricity. It does not emit dust, sulfur compounds, nitrogen and carbon dioxide at all. It is therefore environmentally friendly. (2) Throughout the year it provides consumers with a stable supply of electricity, without which modern civilization could not exist. The annual use of nuclear power plants' capacity exceeds 8,000 h. Moreover, and what is extremely important, the nuclear fuel: uranium, plutonium and thorium will last for many hundreds of years, while coal and gas resources are depleting at an increasing pace. Moreover, after the introduction of a closed fuel cycle with multiple use of nuclear fuel (the so-called nuclear reprocessing), it will be enough for tens of thousands of years. In addition, there are over 4 billion tons of uranium dissolved in seawater, and its technical extraction is under control. Nuclear fuel will therefore last for billions of years. (3) Long, 60-year, lifetime of nuclear power plants. (4) The cost of electricity from nuclear power plants is relatively low, especially after their depreciation—Fig. 5.2. (5) Moreover, the cost of nuclear fuel amounts to merely 5% of the annual operating costs of nuclear power plants. Thus, unlike coal-fired power plants (the cost of coal accounts for approx. 35% of their annual operating costs), and especially gas-fired power plants (the cost of gas in gas and steam power plants reaches up to 75% of the annual operating costs), the cost of nuclear generation is not very sensitive to cyclical changes in the uranium price. Therefore, even a very significant increase in it will cause a slight increase in the price of electricity. It should be strongly emphasized that the unit investment cost of the nuclear power plant shown in Fig. 7.1 will be significantly lower than the cost of a power plant in which the *Clausius-Rankine* cycle is implemented. The unit cost of electricity production will therefore also be significantly lower. Assuming, for example, unit capital expenditures at the level of PLN12/MW (the main component of these expenditures will be the cost of the reactor, while in a conventional system, an equally important component is expenditures on the steam part of a nuclear power plant in which the *Clausius-Rankine* cycle is implemented), the cost of electricity production in the power plant shown in Fig. 7.1 will be significantly lower over the entire range of T years of its operation compared to the costs presented in Fig. 5.2.

R. Bartnik and T. W. Kowalczyk, *Hierarchical Gas-Gas Systems*,
Power Systems, https://doi.org/10.1007/978-3-030-69205-6_7

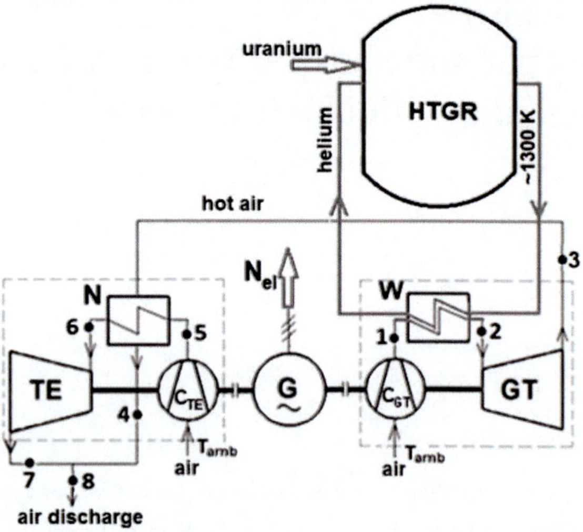

Fig. 7.1 Schematic diagram of a hierarchical *gas-gas* engine with a high-temperature HTGR nuclear reactor

For example, for a standard nuclear power unit lifetime of T = 60 years and for an interest rate of the investment capital of even r = 5%, it will be significantly lower than PLN200/MWh. It should be additionally, what is extremely important, very strongly emphasized that in the case of gas and gas power plants, significant problems and costs related to the water management of the *Clausius-Rankine* steam cycle are "eliminated". Therefore, gas and gas-fired nuclear power plants can be built where there is no water, which should be strongly emphasized once again. The system in Fig. 7.1 will therefore be the source of electricity with the lowest unit generation cost. It will be at least several dozen percent (sic) lower than the costs obtained in all other currently available energy technologies that use fossil fuels, and it will be several times lower compared to the unit cost of renewable generation, [1] (see also Table 5.2).

The thermodynamic "drawback" of the use of nuclear power plants are the water reactors operating in some of them. As a result, the net efficiency of electricity generation is relatively low—it does not exceed 34%. However, due to the development of High Temperature Gas-cooled Reactors (HTGR) which has been going on for several years now, this problem has been solved. Currently, the third generation of these reactors, gas reactors with helium as a coolant for the reactor core and with graphite as a moderator allow the construction of nuclear power plants that realize the *Clausius-Rankine* cycle of a steam turbine with supercritical steam parameters, i.e. with efficiencies of 45%. The helium outlet temperature from the reactor core is already over 1000 °C, therefore obtaining the parameters of live and reheated steam fed to the steam turbine in a conventional non-hierarchical engine, i.e. in an engine where only the *Clausius-Rankine* cycle is implemented at the level of over 600 °C, is no longer a problem. Helium is also characterized by good heat dissipation properties and a small cross-section for neutron capture, and, which is particularly important for nuclear safety reasons, its inactivity.

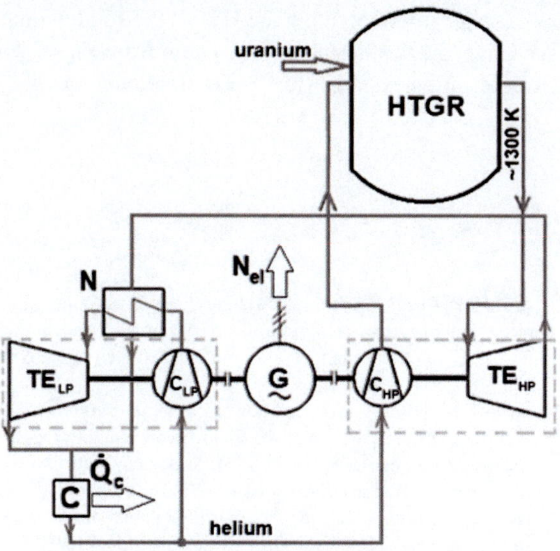

Fig. 7.2 Schematic diagram of a hierarchical gas/gas engine with a high-temperature HTGR nuclear reactor with helium as the only circulating medium

The replacement of natural gas in systems with hierarchical engines, both *gas–gas* and gas-steam ones, Fig. 1.1, with enriched uranium used in HTGR reactors, Fig. 7.1, is therefore also justified due to the above.

It should be noted that in hierarchical gas-steam engines, the optimal pressure of steam produced in the recovery boiler feeding the steam turbine, i.e. the pressure guaranteeing its maximum power, is only a few megapascals, from approx. 5.5 to approx. 10 MPa [2] and not 25–30 MPa, as in a conventional non-hierarchical engine, i.e. in an engine in which realizes the *Clausius-Rankine* cycle with supercritical steam parameters. Therefore, low pressure significantly reduces the capital costs of the steam part of the gas-steam engine plant equipment.

As a result of replacing natural gas with enriched uranium in the engine system shown in Fig. 7.1, there is no gas combustion chamber KS in the system (Figs. 3.2 and 4.1). It is replaced by a indirect-contact heat exchanger W, in which the helium exhaust from the HTGR reactor with a temperature of over 1000 °C heats the air downstream of the compressor. In fact, we are dealing not with a gas turbine, but with a much cheaper turboexpander. Were the gas turbine replaced with a helium turboexpander directly included in the cooling cycle of the HTGR reactor, this would further simplify the engine system and—due to avoiding the loss of exergy in the heat exchanger W caused by the irreversibility of the heat flow from helium to the heated air (formula (2.7))—increase its efficiency by several percentage points above 38% ($\eta_{G-G} = 38\%$ for $T_2 = 1300$ K, Figs. 3.9 and 4.23). It would be even more advantageous if helium was the only circulating medium in the hierarchical *gas–gas* engine system—Fig. 7.2.

In order to raise the economic indicators of a nuclear power plant, work is being carried out not only to increase the thermal efficiency at partial and minimum load [3], but also to increase the power factor of the power plant [4]. This can be achieved, for example, by integrating the HTGR thermodynamic cycle with a SOE electrolyze (see Chap. 5). Furthermore, the system from Figs. 7.2 can also be used in municipal district heating systems (Fig. 3.3) using reactors with relatively low thermal powers [5].

References

1. Bartnik R, Buryn Z, Hnydiuk-Stefan A (2017) Methodology and a continuous time mathematical model for selecting the optimum capacity of a heat accumulator integrated with a CHP plant (in Polish: Ekonomika Energetyki w modelach matematycznych z czasem ciągłym). Wydawnictwo Naukowe PWN, Warszawa
2. Bartnik R (2009) Gas-steam power plants and combined heat and power plants. Energy and economic efficiency (in Polish: Elektrownie i elektrociepłownie gazowoparowe. Efektywność energetyczna i ekonomiczna), WNT, Warszawa (reprint 2012, 2017).
3. Kowalczyk T, Badur J, Ziółkowski P (2020) Comparative study of a bottoming SRC and ORC for Joule-Brayton cycle cooling modular HTR exergy losses, fluid-flow machinery main dimensions, and partial loads. Energy 206:118072. https://doi.org/10.1016/j.energy.2020.118072
4. Kowalczyk T, Badur J, Bryk M (2019) Energy and exergy analysis of hydrogen production combined with electric energy generation in a nuclear cogeneration cycle. Energy Convers Manage 198:111805. https://doi.org/10.1016/j.enconman.2019.111805
5. Szargut J, Ziębik A (2007) Combined heat and power—thermal power plants, in: Wydawnictwo Pracowni Komputerowej Jacka Skalmierskiego, Katowice–Gliwice

Index

R. Bartnik and T. W. Kowalczyk, *Hierarchical Gas-Gas Systems*, Power Systems, https://doi.org/10.1007/978-3-030-69205-6